A CURIOUS GARDEN OF HERBS

A Curious Garden *of* Herbs

Cultivated and Wild; Culinary, Medicinal, Cordial, and Amusing; of the Eighteenth-Century Southern Frontier

KAY K. MOSS AND SUZANNE S. SIMMONS

The University of Georgia Press *Athens*

A Wormsloe
FOUNDATION
PUBLICATION

Athens, Georgia 30602
www.ugapress.org

Designed by Erin Kirk
Set in 10/14 Adobe Caslon Pro
Printed and bound by Versa Press
The paper in this book meets the guidelines for permanence and durability of the Committee on Production Guidelines for Book Longevity of the Council on Library Resources.

Most University of Georgia Press titles are available from popular e-book vendors.

Printed in the United States of America
24 23 22 21 20 C 5 4 3 2 1

Library of Congress Cataloging-in-Publication Data

Names: Moss, Kay, author. | Simmons, Suzanne S., 1955– author.
Title: A curious garden of herbs : cultivated and wild; culinary, medicinal, cordial, and amusing; of the eighteenth-century southern frontier / Kay K. Moss and Suzanne S. Simmons.
Other titles: Publications (Wormsloe Foundation)
Description: Athens : The University of Georgia Press, [2020] | Series: Wormsloe foundation publication | Includes bibliographical references and index.
Identifiers: LCCN 2020013178 | ISBN 9780820357829 (hardback)
Subjects: LCSH: Gardens—Southern States—History—18th century. | Kitchen gardens—Southern States—History—18th century. | Herb gardening—Southern States—History—18th century.
Classification: LCC SB453.2.A13 M677 2020 | DDC 635.0975—dc23
LC record available at https://lccn.loc.gov/2020013178

CONTENTS

ILLUSTRATIONS

FIG. 1. Heartsease

John Gerard, *The Herball*, 1636

PREFACE

Heartsease—such a lovely name. So appropriate. Just the sight of its gentle lavender and yellow blooms brings joy to the spirit and ease to the heart.

It's late March, a bit early for the garden season, yet the kitchen garden is already awakening. Bright green parsley stands in showy contrast to the slender needles of the chives. Sage and melissa peek through the woven fence, ripe for the picking, crowding to catch the sunlight shining warm upon my shoulders as I survey the garden.

Beneath the sleeping musk rose, common violets bloom, spilling splashes of purple amongst the thyme. The damask rose has begun to bud. Its deep pink flowers will soon blow—if a late-season freeze doesn't interfere. Damask likes the cool weather. Soon we'll be infusing a new batch of rose brandy for flavoring puddings and cakes. We shall also collect rose petals to make comfits; those do help ease a dose of bitter medicine.

In the back corner, hop poles stand ready. The radish bed is thriving, kept company by a few errant carrots and parsnips. Young spinage looks promising among tender spears of sparrow grass, just now sending forth. I'm envisioning sparrow grass rolls for Easter. Those will be a fetching accompaniment to Grandmother's traditional herb pudding. A bevy of wildings creep into the garden and around the dooryard—woodsorrel, dandelion, and winter cress—each will add flavor to our eagerly anticipated springtime meal, the hen's first eggs of the season combined with those health-filled herbs to thin the blood!

In the center of it all stands the monarch of the garden, rosemary. Planted long ago by the hands of a dear friend, this stately shrub has endured many seasons, prospering along with the family's generations. Each time I clip a sprig, I remember other branches from this very rosemary that lovingly adorned weddings, marked the passing of loved ones, and celebrated festive winter gatherings. Now showing age, our resilient rosemary is nonetheless strong and vibrant, bedecked with lavender blooms promising more years ahead.

A flurry of apple blossoms swirling about adds a sweet scent to the breeze today. There is another familiar scent wafting. No wonder—I've been absentmindedly fondling a sprig of winter marjoram as I daydream about the health, the flavors, and the joys this garden promises, and the daily care.

Well, enough dallying. 'Tis a fine sunny day and much to be done—and despite the toil, my heart's ease is here.

How Does Your Kitchen Garden Grow?

Which herbs might *you* invite into your kitchen garden? And why?

Each plant suggests a story about the gardener. Cayenne and horseradish say that you enjoy zesty sauces. Abundance of chamomile and mint hint that you may be prone to tummyaches. One might puzzle over odd clumps of purslane and violets. Are these wild herbs bound for your salad bowl, or are they simply invasive (though useful) weeds of a relaxed gardener?

Just as your garden plan reflects your interests and tastes, this present work reflects the interests and passions of the authors. The old traditions infuse our everyday lives.

We began acquiring basic expertise toward this book as youngsters roaming local woods and fields. An important lesson was learned by savoring luscious fully ripened persimmons, and another by biting into an underripe one. The joys of foraging for wild edibles continued into adulthood and cumulated in organizing wild foods festivals at the Schiele Museum in Gastonia, North Carolina, and leading field trips and wild foods programs at other museums, parks, and Girl Scout camps.

This book was conceived twenty-six years ago as a booklet documenting a sampling of herbs appropriate for a modest historical kitchen garden. Enthusiasm for that little volume encouraged our continuing botanical explorations—in the garden, in the library, and in field and forest. Fascinating experiments with herbs, cultivated and wild, have accompanied our four decades spent researching and interpreting the historical landscape and early domestic technologies of the Carolina piedmont.[1]

Need for a balance of plants for medicine, cookery, and other domestic uses guided our collection of gardening lore. We hope you will find new joy in these traditional herbs—and weeds. Perhaps you will discover new favorites to tuck into your garden or inspiration to re-create an eighteenth-century kitchen garden. Perhaps you will gather garnishes for your dinner plate from your backyard—violet leaves and blossoms, flowers of the Judas tree, or even a sprinkling of dandelion petals. Each plant has a story to tell. From gardening guides, cookery books, and domestic medical volumes we draw a unique set of information, dependent on personal experience, interests, or circumstance.

Caution: Please keep in mind that this *Curious Garden of Herbs* is not meant to be a guide for twentieth-century medical use but rather an intriguing history lesson. Some of the plants included in this volume can be quite dangerous. You may be sensitive to a particular plant's toxins, whether through touch or in food or dose.

A CURIOUS GARDEN OF HERBS

Introducing the Kitchen Garden

A good Kitchen Garden is almost as necessary . . . as a Kitchen to the House. —P. Miller, *Gardeners Dictionary*

This collection of herbal fact and lore illuminates the *why* rather than the *how* of the kitchen garden, offering not a how-to of gardening methods but rather tracing herbs and their uses back to earlier times and places. A volume could be written about each of these fascinating herbs.

Practicality and necessity were guiding themes for gardening in eighteenth- and early-nineteenth-century rural and frontier settlements in America. There were plants for food, for seasoning, for medicine, for dye, for repelling insects, and for scent. Many of these plants were decorative as well, but utility was undoubtedly of first consideration. "[Not] that we meane . . . the Kitchin garden should want flowers" (W. Lawson, *Country House-wives Garden*, 76). Our eighteenth-century ancestors likely intermingled herbs, small vegetables, and flowers.

> A kitchen-garden or an orchard is susceptible of intrinsic beauty; and may be so artfully disposed . . . as by variety and contrast to contribute to the beauty of the whole.
> —*1771 Encyclopaedia Britannica*, 647

Today's specialized "herb garden" was not typical in early American settlements. Certainly parsley, sage, rosemary, and thyme were to be expected, along with the basic spinach, savory, marjoram, and chives. However, by the broader definition of "herb," the kitchen garden was likely to have included asparagus, radishes, sweet violets, clove pinks, nasturtiums, even cucumbers. Similarly "vegetable" was defined more broadly as a useful herbaceous plant—nearly synonymous with "herb."

The small farm's dooryard or kitchen garden, managed by the busy housewife, likely contained a limited inventory of common plants. Other useful plants customary to the eighteenth century (as artichoke or saffron) were luxuries rather than staples. Herb choices reflected the refinement and prosperity of the household as well as the locale and maturity of the garden.

> Onions, sallary [celery], leeks, parsley, thyme, and savory, are also necessary in harvest-time, because with these our country housewife cooks up her lean sorts of beef, her pieces of bacon or pork, her offald, cold turnips, carrots, cabbage, or potatoes. And if the meat is a little tainted, yet by her skillful management in the use of some of these roots and herbs, she may recover such meat, by causing it to be hashed or minced according to the art of good housewifery.
> —Ellis, *Country Housewife's Family Companion*, 98

FIG. 2. **Frontier Landscape**

Donegal, Pennsylvania, homestead featuring split rail fence and board fence surrounding kitchen garden and farm animals. Figures in center fence line are obscured by damage along fold line.

Attributed to Nicholas Garrison Jr. (1726–1802)

FIG. 3. **Landscape of a Middling Plantation**

This watercolor offers a rare glimpse of the developing South. The early-nineteenth-century plantation includes a typical farmhouse with later additions. While houses often survive, farm landscapes with outbuildings often do not.

Artist unknown, Virginia–North Carolina line, c. 1835–45

FIG. 4. A View of Salem in North Carolina 1787

Mid-eighteenth-century Moravian settlers followed the Great Wagon Road from Pennsylvania into central North Carolina. This image with the burgeoning village of Salem in the background documents the changing landscape.

Ludwig Gottfried von Redeken, 1787

> Every family, besides a little garden for the few vegetables which they cultivated, had another small enclosure . . . which they called a truck patch, in which they raised corn for roasting ears, pumpkins, squashes, beans and potatoes.
> —Doddridge, *Notes on the Settlement*, 88

That "little garden," the basic kitchen garden of a country housewife, is the focus of this *Curious Garden of Herbs*. Major challenges arise in identifying and locating appropriate species for re-creating a period garden. We must look to today's botanists, seed savers, and historian-gardeners for old species. Period illustrations and descriptions are helpful, though not conclusive.

Settlers brought along starts of favored herbs, trees, and shrubs, or acquired seeds, cuttings, or plants from neighbors and travelers or from one of the few professional seedsmen. Journals and letters from early gardeners tell us of such botanical acquisitions.[1]

Settlers were delighted to find herbs in this strange new land similar to those they had known in their homelands. When encountering unfamiliar plants, pioneering folk were eager to assess their value. Establishing familiarity with the new botanicals was basic to successfully carving out a life in the wilderness. Furthermore, among those people with a bit of leisure, the study of botany and collecting of exotics were popular eighteenth-century avocations. Those interests remain alive among today's herb enthusiasts, cooks, and foragers.

Of course there were many kinds of settlers. We might ponder which vegetables or herbs would have been found growing in a newly established frontier garden or on a hardscrabble farm in contrast to the kitchen garden of a prosperous inland plantation. The garden of the local spinner and dyer would have been decidedly different from that of a domestic medical practitioner. Each garden had a personality reflecting the needs, economics, culture, and interests of the householder.

Publications and personal manuscripts yield details of early American gardens—whether twenty inaccessible, marshy miles from Charleston, two hundred miles into the interior of piedmont Carolina, or across the Appalachians into the Kentucky wilderness.

Frontier Gardeners

Rural and village gardens served the needs of American settlers who lacked access to regular markets and shops. Most of the gardeners who left us records were of the "middling sort"—farmers and townspeople, merchants, traders, and artisans—often aided and influenced by enslaved or indentured workers. There were few wealthy gentry among those settling the early southern frontier.

The concept of "frontier" was constantly changing. The first frontier was the barrier islands. By the eighteenth century, American coastal areas were becoming rather thickly populated. Newly arrived settlers found they had to move farther inland to find affordable land on the expanding frontier. The frontier grew to include the piedmont and mountains of Pennsylvania, Virginia, North Carolina, South Carolina, and Georgia; and, as the century progressed, newly opening lands to the west. This southern frontier was alternately referred to as the backcountry, the upcountry, or simply the wilderness.

Those early settlers were of diverse ethnic stock. As they adapted to their American environment they developed distinctive lifeways and artistic expressions based on homeland traditions. The traditions of

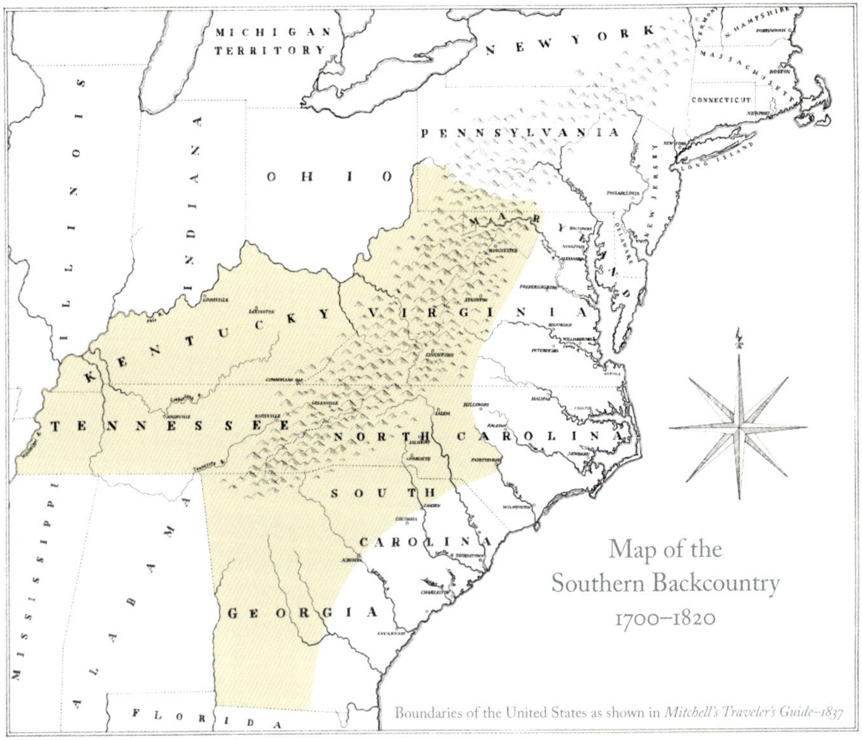

FIG. 5. The Southern Backcountry (1700-1820)

Henry Glassie noted in *Pattern in the Material Folk Culture* that "the [frontier] regions have fuzzy, syncretistic borders and any attempt to define them on a map is a process of constant compromise."

Based on "The Southern Regions" map, Museum of Early Southern Decorative Arts

the numerous Ulster Scots and German settlers were shaped by ideas borrowed from their English, Welsh, and French neighbors with a seasoning of African and Native American influences.

There was a strong cultural similarity between the southern and middle colonies that followed the primary settlement pattern down the Great Wagon Road from Pennsylvania through the Valley of Virginia and the Carolina piedmont into Georgia. A smaller migration merged northwestward from Charleston and Savannah. Along with those streams of people flowed plants, customs, and knowledge. Trade between port cities and inland communities broadened the distribution of plants and information.

Documenting a Period Garden

The works of popular European authors were available to early Americans, as verified through period newspaper advertisements, store ledgers, and personal library inventories. Herbals, cookery books, domestic medical volumes, and gardening guides were among householders' treasured possessions. Books were expensive and largely limited to wealthy colonists, but selected advice from those popular volumes was published in almanacs, newspapers, and "pocket" garden calendars. The nineteenth century ushered in a series of more accessible regional gardening guides especially for the American gardener. Those expensive and rare English and American sources are once again available in affordable reprints or electronically.

Fact and folklore were shared person to person. Journals, letters, and household inventories give us glimpses into specific gardens. The presence of a particular herb in America can be verified through such sources. A variety of personal records have been published and are currently available; others may be awaiting your discovery in the nearest university manuscript collection, your local library or museum, or even a trunk in your attic.

Today we compile, compare, analyze, and comment, carefully documenting each of our sources. As the word *research* indicates, we search again, adding to the body of information and drawing new conclusions.

LANGUAGE AND READABILITY

Eighteenth-century English seems at times a foreign language. We have followed capricious capitalization and punctuation of period sources, except as noted. Original spelling has been maintained with the exception of converting to modern form the old-style long *s* and the overlapping usage of *i* and *j* as well as of *u* and *v* of the archaic twenty-four-letter alphabet (see figure 6). The relic abbreviations *y*[e] (the) and *y*[m] (them) have been transcribed to today's preferred form (as they would have been pronounced then and are now pronounced). Square brackets alert the reader to inserted or corrected words. The use of italics has been modified for readability. A small number of *sic* messages have been used where misunderstanding might arise.

Retaining the charm and flavor of the wording of early writers, cooks, and gardeners slips us into their mindset. The old sources also reveal folklore and unusual gardening hints too useful and entertaining to ignore, such as:

> Birds are frighted away, with feathers stuck in the ground, or made to play in the wind on a line. Poultrie & Catts are to be hinder'd from scraping & basking by laying brambles, & holy-bushes [holly] on the beds.
>
> —Evelyn, *Directions for the Gardiner*, 111

More important, presenting the original herbal notes allows today's reader to draw fresh interpretations. The language is often open to more than one interpretation. See the Glossary for historical definitions.

INTERPRETING SOURCES—MORE THAN MEETS THE EYE

The language and grammar of early authors can lead to confusion and misinterpretation, as in a charming potpourri "to Smell unto for Melancholy." At first reading this formula seems straightforward, and we interpreted it literally for many years. We led dozens of workshop participants in creating lovely little bags of this fragrant herb-and-spice mixture during museum programs before we began to question the details.

> ***A Bag to Smell unto for Melancholy, or to cause one to sleep***
> Take drie Rose leaves, keep them close in a glasse which will keep them sweet, then take powder of Mints, powder of Cloves in a grosse powder, and put the same to the

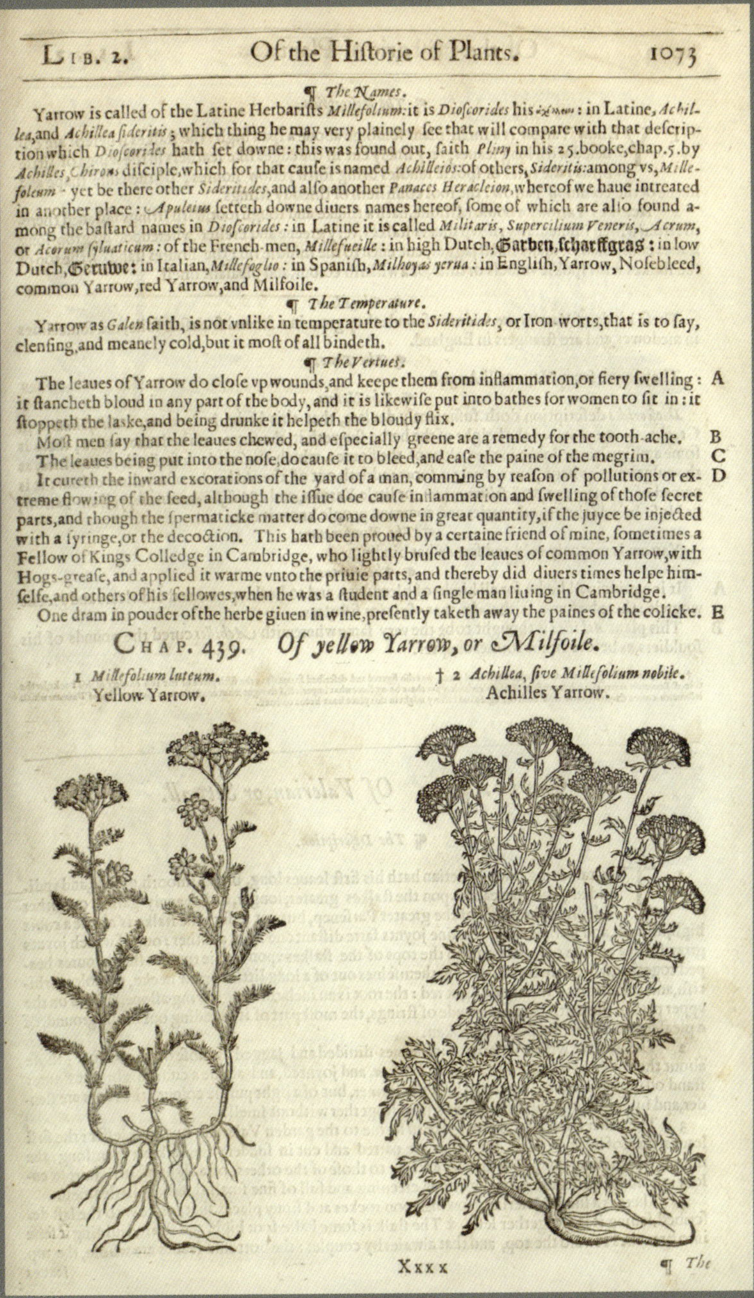

¶ *The Names.*

Yarrow is called of the Latine Herbariſts *Millefolium*: it is *Dioſcorides* his [illegible] : in Latine, *Achillea*, and *Achillea ſideritis*; which thing he may very plainely ſee that will compare with that deſcription which *Dioſcorides* hath ſet downe: this was found out, ſaith *Pliny* in his 25. booke, chap. 5. by *Achilles*, *Chirons* diſciple, which for that cauſe is named *Achilleios*: of others, *Sideritis*: among vs, *Millefoleum* · yet be there other *Sideritides*, and alſo another *Panaces Heracleion*, whereof we haue intreated in another place: *Apuleius* ſetteth downe diuers names hereof, ſome of which are alſo found among the baſtard names in *Dioſcorides*: in Latine it is called *Militaris*, *Supercilium Veneris*, *Acrum*, or *Acorum ſyluaticum*: of the French-men, *Millefueille*: in high Dutch, **Garben, ſcharſſgras**: in low Dutch, **Geruwe**: in Italian, *Millefoglio*: in Spaniſh, *Milhoyas yerua*: in Engliſh, Yarrow, Noſebleed, common Yarrow, red Yarrow, and Milfoile.

¶ *The Temperature.*

Yarrow as *Galen* ſaith, is not vnlike in temperature to the *Sideritides*, or Iron-worts, that is to ſay, clenſing, and meanely cold, but it moſt of all bindeth.

¶ *The Vertues.*

The leaues of Yarrow do cloſe vp wounds, and keepe them from inflammation, or fiery ſwelling: it ſtancheth bloud in any part of the body, and it is likewiſe put into bathes for women to ſit in: it ſtoppeth the laske, and being drunke it helpeth the bloudy flix. A

Moſt men ſay that the leaues chewed, and eſpecially greene are a remedy for the tooth-ache. B

The leaues being put into the noſe, do cauſe it to bleed, and eaſe the paine of the megrim. C

It cureth the inward excorations of the yard of a man, comming by reaſon of pollutions or extreme flowing of the ſeed, although the iſſue doe cauſe inflammation and ſwelling of thoſe ſecret parts, and though the ſpermaticke matter do come downe in great quantity, if the juyce be injected with a ſyringe, or the decoction. This hath been proued by a certaine friend of mine, ſometimes a Fellow of Kings Colledge in Cambridge, who lightly bruſed the leaues of common Yarrow, with Hogs-greaſe, and applied it warme vnto the priuie parts, and thereby did diuers times helpe himſelfe, and others of his fellowes, when he was a ſtudent and a ſingle man liuing in Cambridge. D

One dram in pouder of the herbe giuen in wine, preſently taketh away the paines of the colicke. E

Chap. 439. *Of yellow Yarrow, or Milfoile.*

1 *Millefolium luteum.* Yellow Yarrow.

† 2 *Achillea, ſiue Millefolium nobile.* Achilles Yarrow.

FIG. 6. Yarrow

This image from a popular seventeenth-century English herbal highlights the flavor of the language.

John Gerard, *The Herball*, 1636

FIG. 7. Sassafras

This is a fine example of attention to detail in eighteenth-century botanical illustration.

Elizabeth Blackwell, *A Curious Herbal*, 1771 ed.

Rose leaves, then put all these together in a bag, and take that to bed with you, and it will cause you to sleep, and it is good to smell unto at other times.

—*Ram's Little Dodoen* (1606), 280[2]

What a delightful early aromatherapy. Although it seems straightforward, closer examination reveals a dozen possible combinations of rose leaves, mints, and cloves that fit this formula.

In period sources "rose leaves" generally refers to petals, and this interpretation seems reasonable, even most likely in this context. However, certain rose foliage—that of the eglantine or sweet briar—is also fragrant.

As for the mints, either spearmint, peppermint, or orange mint would be a period-appropriate choice. Each would contribute a signature scent to the blend.

Cloves seems at first obvious. Clove was already a well-known spice in eighteenth-century Europe and America through trade with the Spice Islands. The clove still popular in the twenty-first century is in reality a flower bud from an Indonesian tree. However, as we delve more deeply into the language of the period sources we find that *clove* appears as an alternate name for the fragrant dianthus, clove gillyflower or clove pink. Was the ancient herbalist suggesting buds from an exotic tree or fragrant flower petals from the garden? Do we stir dried dianthus petals or spicy clove buds into our "bag to smell unto"? Either would certainly be a pleasant ingredient for a scented pillow or sachet.

Even the original source of this treatment for melancholy is vague. Although *Ram's Little Dodoen* was based on Rembert Dodoens' *New Herbal* of 1577, the above formula is nowhere to be found in that volume. Such is the fascinating yet curious adventure of historical interpretation.

BOTANICAL ILLUSTRATION—THROUGH THE EYE (AND IMAGINATION) OF THE BEHOLDER

Early botanical illustrations aid us in seeking old species, or at least similar ones. Oftentimes the early species no longer exist. Others may be awaiting rediscovery at an overgrown homesite, an abandoned urban cemetery, or a secluded cove deep in the Appalachians. The illustrations included here were familiar to eighteenth-century gardeners who may have had access to the same volumes we consulted during this project, from Dodoens through Green.

The first important female botanical artist, Elizabeth Blackwell, produced *A Curious Herbal, containing five hundred Cuts, of the most useful Plants* . . . in two volumes in 1737 and 1739. The accompanying notes on gardening were copied from Joseph Miller's popular *Botanicum officinale*. Blackwell's well-received publication gave her the funds to obtain her husband's release from debtor's prison and allowed her to support herself throughout her life. Botany was one of the few studies considered appropriate for an eighteenth-century lady, not only in England but also in the colonies.

In examining old volumes of gardening and plant wisdom we encounter charmingly hand-colored illustrations. Those were generally the work of professional artists; however, during the eighteenth and nineteenth centuries, hand coloring floral illustrations was a popular pastime for the gentry. There was even the rare sophisticated coloring book such as Robert Sayer's 1760 *The Florist*, "intended for the use and amusement of

Gentlemen and Ladies Delighting in that Art," complete with advice on paints. Into the following century a variety of lovely botanical volumes were published with illustrations ready to be colored by those so inclined.

Perhaps you will follow tradition and choose to color the illustrations in your copy of this *Curious Garden of Herbs*. These historical images beckon to anyone wishing to take paints or colored pencils to the garden to study detail: each leaf a different shade of green, stems sporting surprising hues, blossoms tinted with so many lavenders and pinks. There is no more pleasurable way to indelibly engrave plant recognition in the memory.

CAUTION AND COMMON SENSE IN THE GARDEN—SENSES AND SENSIBILITY

The herbal lore collected here brings stories from more than two centuries past into your garden today. The curious gardener, armchair historian, adventurous cook, gourmand, historical interpreter, and folklorist may find new fascination in common herbs, flowers, and even weeds. Every sense comes into play in this garden—intriguing scents, titillating tastes, widely varied textures, ever-changing colors, and the gentle sounds of breeze and insects. A sixth sense must also be exercised: your sense of caution!

> How cautious then ought our Sallet-Gatherers to be, in reading ancient Authors.
>
> —Evelyn, *Acetaria*, 165

We should exercise every caution in experimenting with any new food or drink. Antiquated medical practices and questionable culinary suggestions have been included here not as recommended use, but to round out the histories. A number of these plants are now recognized as poisonous to some degree. For example, the herb tansy, once common in springtime dishes, is now considered a toxic plant.

Be wary of ingesting any unfamiliar herb—among them, but not limited to, pennyroyal, castor, poke, and even sassafras; each contains toxins. The danger is directly proportional to the extent of plant exposure as well as to personal sensitivity.

Be aware that some people experience skin irritation from the very chemicals responsible for fragrance and flavor. Contact with rue, asparagus, and even chamomile or marigold may cause a rash. Be wary and aware that you could be that individual who will react adversely to a given herb. Being green or organic or "natural" does not make an herb safe. "Old fashioned" does not equate with "tried and true."

Medical notions from centuries past must likewise be interpreted as history rather than advice. Long-term serious consequences, including cancer or liver damage, are possible and often go unrecognized. Additional problems arise from misjudging a safe versus a potentially dangerous dosage.

You may enjoy digging out additional bits of historical information by exploring our primary sources. Just as our ancestors selected plants to suit their individual needs, you will be attracted to those plants with lore appealing to your personal experiences and interests. When your choices are solidly based on information available to early frontier folk, your garden will reflect a slice of life in the eighteenth century.

Welcome to our imagined garden! In this collection of historical herb information you will find a balance of

European and American sources from early to later periods. Related culinary and medical practices from the seventeenth century into the early nineteenth century expand and bracket eighteenth-century gardens and plant usage.

CHECKLIST OF PRINCIPAL SOURCES

Herbals

Dodoens (England, 1578)
Parkinson (England, 1629)
Gerard (England, 1633)
Culpeper (England, 1770, little changed from 1653)
Green (England, 1823)

Gardening Guides

Evelyn (*Directions*, England, 1686)
P. Miller (*Dictionary*, 1735; *Kalendar*, England, 1759)
Bradley (*Kalendar*, England, 1720)
Mawe (England, 1773)
Squibb (South Carolina, Georgia, and North Carolina, 1787)
John Randolph (Virginia, 1793)
M'Mahon (mid-Atlantic and southern states, 1806)
Gardiner and Hepburn (northern Virginia, 1826)
Cobbett (New York, 1821 and 1856)

Medical Guides

Quincy (England, 1736)
Tennent (Virginia and Maryland, 1736)
Buchan (England and America, 1774)
Wesley (England and America, 1791)
Thomson (America, 1802)
W. Barton (America, 1810, 1812)
Gunn (Tennessee, 1830)
Howard (Ohio, 1836)

General Household Advice and Cookery Books

Tusser (England, 1557)
W. Lawson (England, 1660)
Evelyn (*Acetaria*, England, 1699)
Bradley (*Country Housewife*, England, 1736)
Smith (England and America, 1742)
Glasse (England, 1747; America, 1804, if so noted)
Ellis (England, 1750)
Raffald (England, 1769)
Loffler (Germany, 1795)
Simmons (America, 1796)
Von Rumohr (Germany, 1822)
Mary Randolph (Virginia, 1824)
Dods (Scotland, 1829)
Child (America, 1832)
Bryan (Kentucky, 1839)
Rutledge (South Carolina, 1847)

Personal Journals and Manuscript Household Memoranda

Unidentified (cookery, etc.) MS (Virginia, 1700)
Pre-1744 (cookery, etc.) MS (Virginia, early 1700s)
W. Lawson (Carolinas, 1709)
Jane Randolph MS (Virginia, 1743)
Byrd (Virginia and North Carolina, 1758)
Records of the Moravians (North Carolina, 1753–1800)
Kalm (mid-Atlantic colonies, 1770)
Allen MS (North Carolina, late 1700s)
F. A. Michaux (southern states, 1801–2)
Lenoir MS (North Carolina, 1801–39)
Thompson MS (Tennessee, 1803–11)

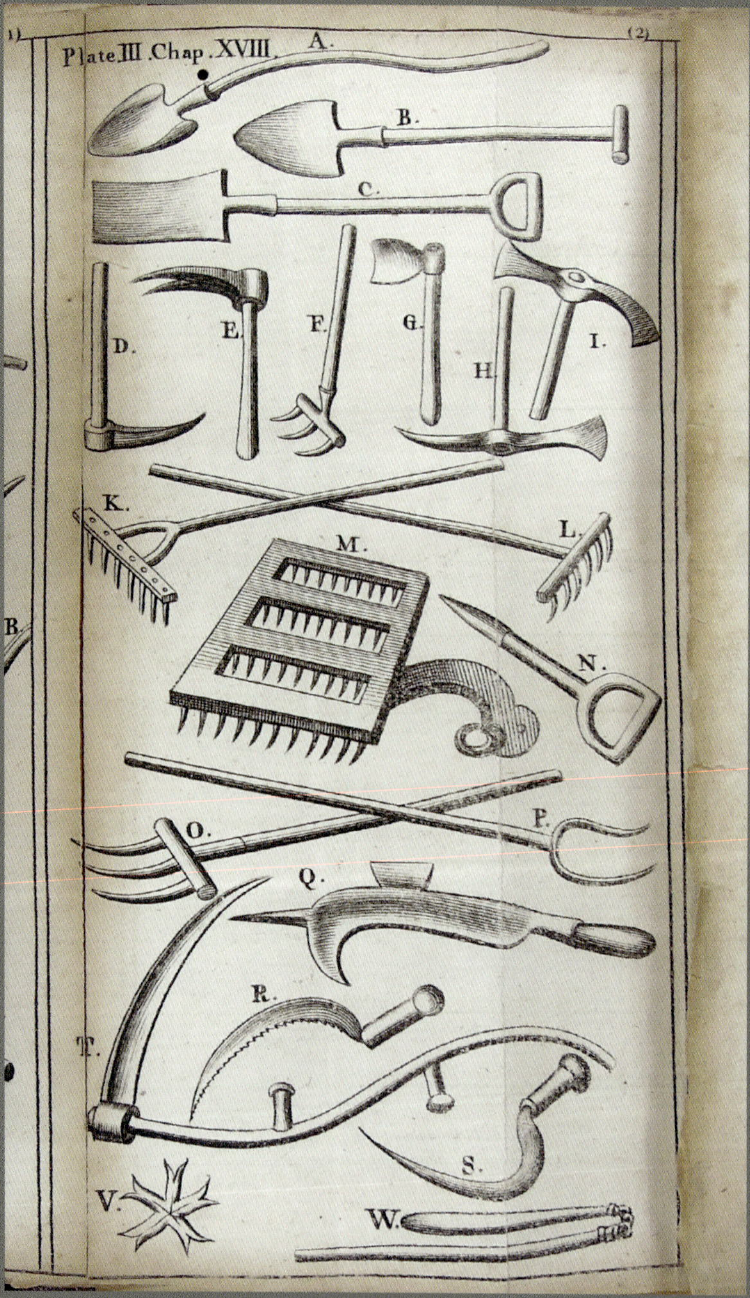

FIG. 8. Garden Tools

A—Spade, B—Common Spade, C—Garden Spade, D—Single Sarr or Pick, E—Double Sarr or Hough, F—Three fang'd Sarr, G—English Sarr or Hough, H—Pickax or Stocking Ax, I—Mattock, K—Wooden Rake, L—Iron Rake, M—Harrow, N—Dibble, O—Prong or Dung-fork, P—Pitch-fork, Q—Falx or Bill, R—Sickle, S—Reaping Hook, T—Sythe, V—Pod of the *Tributus terrestris* Plant, W—Flail.

Richard Bradley, *Survey of Ancient Husbandry and Gardening*, 1725

The Herbal

The Garden Herbs

The Kitchen Garden should include those herbs fitter for the pot and kitchen than for the hand or bosome.

—Parkinson, *Paradisi*, 473

Anise, Aniseed. *Pimpinella anisum*

Anise was one of many plants grown in American gardens in the early eighteenth century, including the North Carolina Moravians' medical garden. In Tusser's 1590 *Good Husbandry* we find anise listed among "Necessary Herbs to grow in the Garden for Physic." Along with dill, fennel, and caraway seeds, aniseed was acclaimed for "windy disorders," as noted by the unknown author of an eighteenth-century South Carolina manuscript of remedies and recipes echoing Gerard's 1633 English herbal:[1]

For fits of windy disorders

Take two pennyworth of sweet Fennel seeds two pennyworth of anniseeds powdered fine give to a grown up person one small tea spoonful in any Liquid.

—Unknown author, Herbal and household book

The seed wasteth and consumeth winde, and is good against belchings . . . allayeth gripings of the belly. . . . Being chewed it makes breath sweet . . . it helpeth the yeoxing or hicket,[2] both when it is drunken or eaten dry.

—Gerard, *Herball*, 1035

It is good for opening the pipes, and it is used in Comfits.

—W. Lawson, *Country House-wives Garden*, 78

Outside of comfits and confections, we saw no anise in the American, English, and French culinary recipes we examined. However, German bakers of the period were fond of aniseed in certain breads and little cakes, as in the Springerle below.

The Pastrycooks also make great Use of these Seeds in several of their Compositions, to give them an aromatick Taste and Smell.

—P. Miller, *Gardeners Dictionary*

Little Springerle in the Usual Manner

A pound of sifted sugar is stirred for an hour with 4 eggs. The zest from a lemon is grated and this, along with a pound of fine flour is mixed into the sugar. Then one takes a little piece from the dough . . . rolls it out to the thickness of the back of a thick knife. . . . [A design is then stamped into the dough with a wooden mold.] The baking sheet is spread with butter, sprinkled with anise, and the Springerle is laid on it. They can lie on the baking sheet for a couple of hours, or even over night. Afterwards, one bakes them in an oven which is no longer hot. They must be more white than brown.

—Loffler, *Economical Handbook*, in Roberts, *The Raised Hearth*, 132

Artichoke, Globe Artichoke. *Cynara scolymus*, *Cynara cardunculus* var. *scolymus*

Cobbett's *American Gardener* declared artichoke a "plant little cultivated in America, but very well worthy of cultivation." Although not common, the artichoke was prized as a gourmet addition to a well-spread table. This sophisticated vegetable even found its way into some frontier gardens; artichoke seeds were tucked into an extensive order of seeds and plants bound from Philadelphia to the Illinois country in 1768.[3]

Cookery books on both sides of the Atlantic offered methods for preparing the artichoke buds and blossoms (and the similarly flavored, though unrelated, Jerusalem artichoke). "The best Sort is what the Gardeners call the Globe Artichoke. . . . [The] fleshy Part at the Bottom of the Scales is very thick, therefore is much preferred."[4]

A pre-1744 Virginia personal cookery manuscript included directions for pickling "Harty Choak Bottoms" along with how to dry them. *The Carolina Housewife* suggested serving artichoke bottoms in a simple cream sauce, while *The Virginia Housewife* gave directions for serving the "leaves" accompanied by individual cups of melted butter, a preparation still popular with artichoke lovers two centuries later.[5]

> The parts that are eaten are, the lower end of the thick leaves that envelop the seed, and the bottom out of which those leaves grow. The whole of the head, before the bloom begins to appear, is boiled, the pod leaves are pulled off by the eater, one or two at a time, and dipped in butter, with a little pepper and salt, the mealy part is stripped off by the teeth, and the rest of the leaf put aside, as we do the stem of the asparagus. The bottom, when all the leaves are thus disposed of, is eaten with knife and fork.
>
> —Cobbett, *American Gardener*, 89

> ***To dress Artichokes***
>
> Wring off the Stalks, and put them into the Water cold with the Tops downwards, that all the Dust and Sand may boil out. When the Water boils, an Hour and a Half will do them.
>
> —Glasse, *Art of Cookery* (1747), 11

> The Heads being slit in quarters first eaten raw, with Oyl, a little Vinegar, Salt, and Pepper, gratefully recommend a Glass of Wine . . . at the end of Meals. They are likewise, whilst tender and small, fried in fresh Butter crisp with Parsley. But then become a most delicate and excellent Restorative, when full grown, they are boil'd the common way. The Bottoms are also bak'd in Pies, with Marrow, Dates, and other rich Ingredients.
>
> —Evelyn, *Aceteria*, 143

The Frugal Housewife's receipt for artichoke pie does not seem frugal at all. The book, published in both London and Boston (1772), recommended adding artichoke bottoms to "Trout Pie" or "Lamb-Stone [sheep testicle] and Sweet-Bread [pancreas or thymus gland] Pie" as well as stewed with ox palates or pigeons.[6]

> ***Artichoke or Potatoe Pie***
>
> Take artichoke bottoms, season them with a little mace and cinnamon sliced, eight ounces of candied lemon and citron sliced, eringo roots and prunellas a slit [slice] of each, two ounces of barberries, eight ounces of marrow, eight ounces of raisons of the sun stoned, and two ounces of sugar, butter bottom of the pie, and put in all, one with the other, and eight ounces of butter on the top lid; bake it.
>
> —Carter, *Frugal Housewife*, 98–99

FIG. 9. Globe Artichoke

Basilius Besler, *Hortus Eystettensis*, 1613

An Artichoke Pye
Take the bottoms of 6 or 8 artichokes being boyled & sliced season them with sweet spice [cloves, mace, nutmeg, cinnamon, sugar] mix them with the marrow of 3 bones with fruite & preserves [raisins, currants, prunellas, damsons, gooseberries, grapes, and candied citron, lemon chips, and eringo roots . . . add butter and top lid].[7]
—Unidentified cookbook, circa 1700, in Harbury, *Virginia's Cooking Dynasty*, 232

The ultimate in interplanting:

In those Plantations [gardens or fields] where you intend to plant other Things between your Artichokes, you must allow nine or ten Feet between the Rows . . . sow the Ground between with Radishes or Spinage, and plant two Rows of Cauliflowers . . . in May when the Radishes or Spinage are taken off, then sow a Row of Cucumbers for Pickling . . . and between the Rows of Cauliflowers and the Artichokes, plant a Row of Cabbages or Savoys for Winter Use, which, when the Cauliflowers are drawn off, and the Artichokes gathered, will have full Liberty to grow; and by this Means the Ground is fully employed through the whole Season.
—P. Miller, *Gardeners Dictionary*

Squibb's gardening guide for the southern colonies continued similar advice for the springtime: plant a row of cabbage between artichokes "or sow radishes, spinage and lettuce."

[Artichoke] leaves and stalks enter into many of the diuretic decoctions used by the country people.
—*Edinburgh Dispensatory*, 172

Some plants are also used as rennet . . . [such as the flower buds of] an acrid species of thistle and artichoke.
—Dods, *Cook and Housewife's Manual*, 505

Adventurous gardeners or cooks might also have planted a smaller member of the artichoke genus, the cardoon. "Cardoons [*Cynara cardunculus*] a variety of artichoke of which the stalks are eaten. . . . Their characteristic bitter flavour is not, however, to everyone's taste and this has led to the practice of boiling them initially."[8]

Cardoon Artichoke has been a long time used for culinary purposes, such as for sallads, soups, stewing, &c. The stalks of the leaves being thick, fleshy, and crisp, are the eatable parts, being first blanched by landing them up like celery, to two or three feet high, to render them white, tender, and of an agreeable flavour, which otherwise would be rank and bitter: they are in perfection in autumn and winter.
—M'Mahon, *American Gardener's Calendar*, 197

[Blanching by "earthing" or "landing":] When the plants arrive at their full growth, or between three and four feet high, they should be earthed up almost to their tops, first tying the leaves of each plant close together with hay or straw bands.
—Mawe, *Every Man His Own Gardener*, 250

See also **Jerusalem Artichoke**.

Artemisias.
See **Southernwood**, **Tarragon**, and **Wormwood**.

Asparagus, Sparagus, Sparrow Grass, Sperage. *Asparagus officinalis*

In 1709, English explorer John Lawson noted that asparagus "thrives to a Miracle" in Carolina. Swedish botanist Peter Kalm "frequently saw asparagus growing

FIG. 10. Sparagus, Asparagus

Elizabeth Blackwell, *A Curious Herbal*, 1739

near fences in loose soil on uncultivated fields. It was also plentiful between stands of corn. . . . I have likewise seen it growing wild in other parts" of the middle colonies (1751). This vegetable was a welcomed addition to the early spring diet.[9]

> The sweetest Asparagus is that which grows upon the richest Ground . . . these beds were every Year dunged extremely thick.
>
> —P. Miller, *Gardeners Dictionary*

Interplanting was recommended for the asparagus bed as for artichokes. In Miller's *Dictionary* from 1735 England we find "sow thereon a small Crop of Onions, which will not hurt your Asparagus." Fifty years later, Squibb's 1787 *Gardener's Calendar for South Carolina, Georgia, and North-Carolina* echoed this plan: "It is the custom with some people, who are obliged to make the most of every spot of ground, to sow a thin crop of onions, the first year, on new asparagus beds"; or "you may sow thereon some radish, spinage, or lettuce seed."

> When the buds [shoots] are about three or four inches high, they should be then cut, observing to slip the knife [perhaps a specialized blunt and serrated asparagus knife] down close to the bud, and cut it off about three or four inches within the ground, taking great care not to wound or break off any young bud coming up near it from the same roots.
>
> —Mawe, *Every Man His Own Gardener*, 98

There was wide agreement on the proper method for preparing and serving asparagus. *The Virginia Housewife* and *The Kentucky Housewife* were in complete accord with popular English cookery books such as Eliza Smith's:[10]

> *To Dress Asparagus*
> Let all the stalks be carefully scraped till they look white, cut them of an equal length, and throw them into water; set your stew-pan with a proper quantity of water, having some salt in it, on the fire, and when the water boils put in your asparagus, after being tied up in small bundles. When they are enough . . . take them up, for if they boil too long, they will lose both their colour and taste.
>
> —Smith, *Compleat Housewife*, 15

> The first shootes or heads of Asparagus are a Sallet of as much esteeme with all sorts of persons, as any other whatsoever, being boyled tender and eaten with butter, vinegar, and pepper, or oyle and vinegar, or as every ones manner doth please.
>
> —Parkinson, *Paradisi*, 503

When asparagus is tied in bundles as directed above, it should be set into the boiling water with stems down and tips above the water so that the tender tips gently steam while the stalks cook thoroughly. Asparagus was typically served on toasts dipped in the asparagus cooking liquid and doused with melted butter.

> [The rationale for the following Old World medical advice may be obvious:] They [asparagus] somewhat provoke urine, are good for the kidneys and bladder . . . they are thought to increase seed, and stir up lust.
>
> —Gerard, *Herball*, 1112

Balm, Melissa, Lemon Balm, Baum

Melissa officinalis

The popular 1735 English *Gardeners Dictionary* tells us melissa was "cultivated in Gardens for Medicinal and

Culinary Use." Many authors through nearly two centuries touted the delight and the perceived benefit. The early significance of balm, especially in beverages, crossed the ocean to America. "Balm when acidulated with juice of lemons makes a very pleasant drink." Whether paired with lemon in a tea or simply a fresh sprig in a glass of wine, balm contributes a thoroughly pleasing flavor, and what a bonus if it may indeed "powerfully chase away Melancholy."[11]

> [Melissa or balm is] hot and dry, Cordial and exhilarating, sovereign for the Brain, strengthning the Memory, and powerfully chasing away Melancholy. The tender Leaves are us'd in Composition with other Herbs; and the Sprigs fresh gather'd, put into Wine or other Drinks, during the heat of Summer, give it a marvellous quickness.
>
> —Evelyn, *Aceteria*, 144

> Let a Syrup made with the Juice of it and Sugar . . . be kept in every Gentlewoman's House, to relieve the weak Stomachs and sick Bodies of their poor sickly Neighbors.
>
> —Culpeper, *English Physician*, 27

> Bawm comforts the Stomach, removes Melancholy, cheers the Heart, expels Poison, cures the Plague, Fevers, the Biting of Mad Dogs, and all venomous Creatures. . . . The Leaves dry'd and use'd as Tea is very good for weakly Persons.
>
> —South Carolina and Georgia Almanack for 1758.[12]

> A good breast tea [for chest complaints].
>
> —Fries, *Records of the Moravians*, 2: 575

> Balm is also, a garden herb, and affords a pleasant tea to be drank in fevers.
>
> —Gunn, *Domestic Medicine*, 385

Basil.

Sweet Basil. *Ocimum basilicum*

Bush Basil. *Ocimum basilicum* 'Minimum'

You will rarely find basil in eighteenth-century receipts, making it unlikely to have been among the sweet herbs (parsley, thyme, marjoram, savory, etc.) cultivated in backcountry gardens. However, apparently some people in the colonies did enjoy basil, for we read in *The Gardener's Calendar for South Carolina, Georgia and North-Carolina* that basil was, "in some families, used as a kitchen herb." Bryan's *Kentucky Housewife* was a rare source recommending basil among sweet herbs in a few basic receipts for meats: "Savory Beef," "To Roast a Hind Quarter of Lamb," and "Savory Venison."[13]

> Basil is a very sweet annual pot-herb . . . two sorts. . . . The bunches may be dried for winter use.
>
> —Cobbett, *American Gardener*, 94

> Basil . . . imparts a grateful Flavour, if not too strong, somewhat offensive to the Eyes; and therefore the tender Tops to be very sparingly us'd in our Sallet.
>
> —Evelyn, *Aceteria*, 14

While older herbals had proposed basil for its scent and medical uses, the medical profession of the day downplayed its usefulness. The 1736 *English Dispensatory* declared basil "an ill-scented Plant . . . found in very few Compositions of Medicines."[14]

> The first sort [common basil] is what the college of Physicians have prescribed for Medicinal Use, but [bush basil] is most esteemed for its Beauty and Scent, by those who cultivate them to adorn their Gardens.
>
> —P. Miller, *Gardeners Dictionary*

The ordinary Basil is in a manner wholly spent to make sweet, or washing waters, among other sweet herbs, yet sometimes it is put into nosegayes. The Physicall properties are, to procure a cheerefull and merry heart, whereunto the seede is chiefly used in pouder, &c. and is most used to that, and to no other purpose.
—Parkinson, *Paradisi*, 450

The herbe brused with vineger holden to the nose of suche as are faynt . . . bringeth them againe to theselves.
—Dodoens, *Niewe Herball*, 241

[Basil was thought to] carry a kind of virulent Quality with it. Being applied to the Place bitten by venomous Beasts, or stung by a Wasp, or Hornet, it speedily draws the Poison to it. . . . [T]his Herb and Rue will not grow together, no, nor near one another; and we know Rue is as great an Enemy to Poison as any grows.
—Culpeper, *English Physician*, 31

Betony. *Stachys officinalis*

Considered a "precious herb," betony was recommended for almost all internal or external complaints from injury or disease. It was said to preserve the body "from Witchcrafts also." In the early American South, personal memoranda included betony among a long list of other herbs in polypharmaceutical ointments and draughts such as "Aquae Mirabilis" and "Sady Allens Water." Apart from medical custom, the leaves and flowers were harvested for "Broth, Drink, or Meat" or to be dried or preserved. The professional medical community downplayed the value of this herb, although some domestic practitioners continued to follow advice from early herbals.[15]

[Betony] Leaves and Flowers by their sweet and spicy Taste, are comfortable both in Meat and Medicine. . . . [I]t is a very precious Herb, that is certain, and most fitting to be kept in a Man's House, both in Syrup, Conserve, Oil, Ointment, and Plaster. The Flowers are usually conserved.
—Culpeper, *English Physician*, 41

[Betony] maketh a man to have a good stomack and appetite to his meate: it prevaileth against sower belchings. . . . There is a Conserve made of the floures and sugar good for many things, and especially for the head-ache.
—Gerard, *Herball*, 714–15

Though formerly much celebrated, [Betony] is now [1812] omitted in the catalogues of the British dispensatories; and, in my opinion, very justly.
—W. Barton, *Cullen's Treatise*, 2: 103

Clouds Flying Before the Eye [likely "floaters"]
Take a drachm of powered betony every morning.
—Wesley, *Primitive Remedies*, 61

To cure the toothache . . . put a leaf of betony, up the nose.
—Wesley, *Primitive Remedies*, 119

Borage. *Borago officinalis*

The extraordinary blue of the borage blossoms in the garden may help "expel Pensiveness and Melancholy" even before they are harvested and processed into conserve or cordial. Borage was documented in America in 1709 by explorer John Lawson and at midcentury by the North Carolina Moravians.

Leaves, Flower, and Seed, all or any of them, are good to expel Pensiveness and Melancholy. . . . The Flowers candied or made into a Conserve . . . are chiefly used

FIG. 11. Garden Borage

John Gerard, *The Herball*, 1633

as a Cordial, and are good for those that are weak in long Sickness, and to comfort the Heart and Spirits of those that are in a Consumption, or troubled with often Swoonings, or Passions of the Heart.

—Culpeper, *English Physician*, 50–51

Flowers are used in Medicine, and the Herb for cool Tankards in Summer.

—P. Miller, *Gardeners Dictionary*

Sprigs in Wine, like those of Baum, are of known Virtue to revive the Hypochondriac.

—Evelyn, *Aceteria*, 145

The old verse concerning Borage doth testifie: I Borage bring alwaies courage. . . . Those of our time do use the floures in sallads, to exhilerate and make the mind glad.

—Gerard, *Herball*, 797

Chopped borage, dressed with oil and vinegar, is served as a garnish to boiled dishes.

—Von Rumohr, *Essence of Cookery*, 164

Burrage and Buglosse . . . exceeding good Pot-herbs, good for Bees, and most comfortable for the heart and stomack.

—W. Lawson, *Country House-wives Garden*, 78

Bugloss. *Anchusa officinalis*

Bugloss and Borage were often grouped together as having similar attributes. They were considered "cordial" flowers along with roses and violets and were thought to have "exhilarating" virtues. "Bugloss . . . much like Borrage. . . . The Flowers of both, with the intire Plant, greatly restorative, being Conserv'd."[16]

Bugloss has a slimy sweetish taste. . . . The flowers are one of the four called cordial flowers: the only quality they have that can intitle them to this appellation, is, that they moderately cool and soften without offending the palate or stomach.

—*Edinburgh Dispensatory*, 152

Bugloss and borage were common herbal ingredients in waters, cordials, and conserves intended for medical use. These concoctions combined long lists of herbs and even incorporated such unsavory additions as snails or dung into doses recommended for a range of ills.[17]

Burnet. *Sanguisorba minor*

Burnet is one of many herbs proposed for enhancing a glass of wine. A "fresh sprig in Wine, recommends it to us as its most genuine Element." Although the old herbals considered it beneficial against depression and other ailments, by the end of the eighteenth century Quincy's *Dispensatory* reported burnet "almost wore out of Practice."[18]

A most precious Herb, little inferior to Betony. . . . Two or three of the stalks, with Leaves put into a Cup of Wine, especially Claret, are known to quicken the Spirits, refresh and clear the Heart, and drive away Melancholy.

—Culpeper, *English Physician*, 63

Pleasant to be eaten in sallads, in which it is thought to make the heart merry and glad.

—Gerard, *Herball*, 1046

A cooling tea in cases of bleeding piles.

—Tennent, *Every Man His Own Doctor*, 44

For a Sore Brest

Take Cream, Camamile Large Lavandar, Howlick [houseleek] . . . and Burnet . . . stude [stewed] all Together

FIG. 12. Garden Burnet

John Gerard, *The Herball*, 1633

for the Space of 2 or 3 Aurars [hours] Tele it becomes a Save.

—Blount, Pocket memorandum, 1764–67, N.C. State Archives

[Burnet] takes Freckles off the Face, eases the Tooth-ach, and ripens Swellings. But notwithstanding these wonderful Virtues . . . it is almost wore out of Practice, and now in no great esteem.

—Quincy, *English Dispensatory*, 99

Neither its sensible qualities nor any analysis can make us think of it as a medicine of any value.

—W. Barton, *Cullen's Treatise*, 2: 114

Camomile, Chamomile. Roman, English, or Garden Camomile.

Chamaemelum nobile

German Camomile. *Matricaria chamomilla*

John Lawson, traveling through the Carolinas, commented that camomile "must be kept in the Shade, otherwise it will not thrive." Nevertheless, camomile was listed by gardeners from the Carolina lowcountry to the Illinois frontier. Climate differences must be considered across that broad geographical area and through a century of records. Camomile favored a historically cooler climate.[19]

Falstaff, in Shakespeare's *Henry IV*, alludes to paths carpeted with camomile: "For though the camomile, the more it is trodden on the faster it grows, yet youth, the more it is wasted the sooner it wears." This concept of sitting or walking on a carpet of camomile crossed the Atlantic, as evidenced by lowcountry plantation mistress Eliza Lucas Pinckney's garden plans in mid-eighteenth-century South Carolina, which included seats of camomile; and by Virginian John Randolph's *Treatise on Gardening* noting camomile used "in making green walks, or edgings." Others suggested "the double-blossomed variety makes a pretty edging for the borders of cottage gardens"; and furthermore, "it is sweet smelling, qualifying head-ach."[20]

In all your Gardens and Orchards, banks and Seats of Camomile, Penny-royall, Dasies and Violets, are seemly and comfortable.

—W. Lawson, *Country House-wives Garden*, 84

Wesley's and Gunn's domestic medical advice in America survived full-blown from the European herbal tradition of Gerard.

The decoction of Cammomill made in wine and drunke, is good against coldnesse in the stomacke, soure belchings, voideth winde and mightily bringeth downe the monethly courses [menstruation].

—Gerard, *Herball*, 756

For pain in the stomach from bad digestion. Take fasting, or in the fit, half a pint of camomile tea.

—Wesley, *Primitive Remedies*, 113

The Chamomile flowers when steeped in old whiskey, or any good spirits, and taken two or three times a day, in moderate quantities, is an excellent medicine to give tone or strength to a weak stomach and restore the appetite: for women given in hysterical complaints, this is a valuable remedy.

—Gunn, *Domestic Medicine*, 382

The following therapy illustrates Wesley's *Primitive Physick* (*Primitive Remedies*) in use on the Tennessee frontier. John Thompson penned a number of Wesley's remedies into a personal memorandum book.

Sore and Swelled Breasts

Boil a handful of camomile and as much mallows in milk and water. Foment with it between two flannels, as hot as can be borne, every twelve hours. It also dissolves any knot or swelling in any part.

—Wesley, *Primitive Remedies*, 36; and also Thompson, Commonplace book, 95

We find a late-eighteenth-century domestic medical practitioner in backcountry North Carolina in agreement with Gerard's herbal: Rachel Allen recommended chamomile enemas to treat kidney stones—a clyster of chamomile flowers, elderflowers, and fennel seed.[21]

Cammomill is good against the collicke and stone; it provoketh urine, and is most singular in Clysters [enemas] which are made against the foresaid diseases.

—Gerard, *Herball*, 756

[In 1794, a man] very ill with clogged bowels . . . put camomile and tobacco in the clyster.

—*Records of the Moravians*, 6: 2524

A thoroughly curious cure involving camomile found its way into the *Raleigh (North Carolina) Register* on July 16, 1833. William Lenoir considered this advice useful enough to copy into his personal medical notebook.

Cure for Consumption [Tuberculosis] from

A young Lady who had been doctered much without effect, she went to a place of retirement where she saw a poor sickly dog every morning go and lick the dew of a Camomile bed which restored his health, and she procured the dew from the same Camomile and drank a small quantity ever morning, & got completely cured.

—Lenoir, Medical memorandum

Caraway *Carum carvi*

Common cakes as well as little cakes (cookies) were varied with the addition of caraway seeds. Popular cookery book author Hannah Glasse suggested flavoring seed cakes with rosewater and perhaps saffron, although the basic seed cake is very nice without them. About one tablespoonful caraway is good in the following receipt from a 1756 North Carolina notebook. Omitting the caraway, this is a basic "pound" cake.

To Make a Light Seed Cake

Take a Pound of Flour well Dried, a Pound of Sugar well Sifted, a Pound of Butter, the yelk [yolk] of Twelve Eggs & the whites of Six and a 1/2 oz Carraway Seeds Mixt well together & Bake it.

—Goelet, Household book

Seed cake was traditionally taken to harvesters in the field or offered for supper with cheese. It was, and is, considered especially nice dipped into ale. We have served seed cake with ale and cheese on numerous occasions to enthusiastic reception.

Gastrointestinal distress was treated with caraway or seeds of related herbs: "Comfits, are very good for the stomacke, they helpe digestion, provoke urine, asswage [assuage] and dissolve all windiness." Caraway continued to be recommended more than a century later in America because "its seed . . . is highly aromatic and grateful to the stomach."[22]

The Seeds which we generally make Comfits of, are Carraways, Coriander and Anise-Seeds; these, when they are cover'd with Sugar, are call'd Comfits.

—Bradley, *Country Housewife*, 2: 45

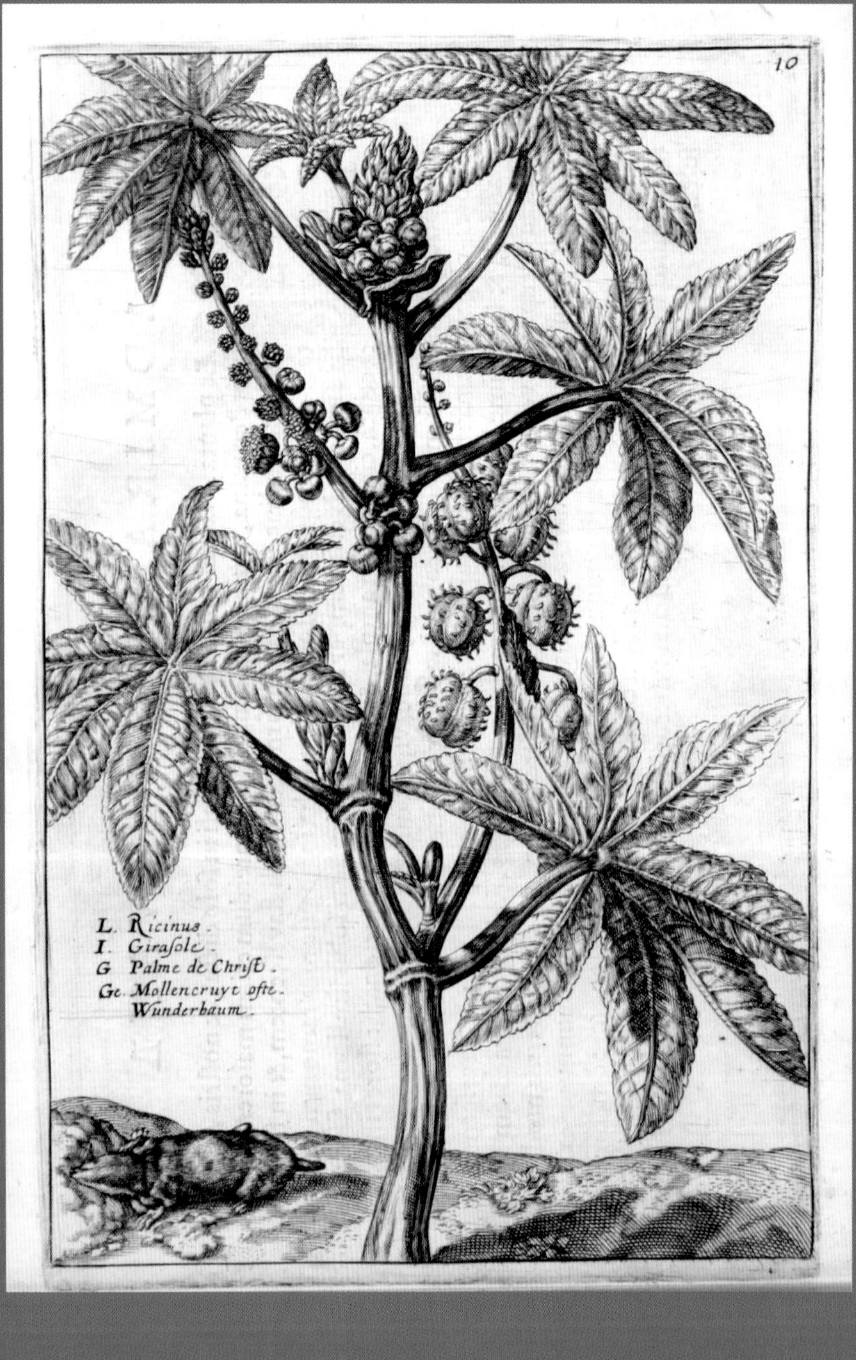

FIG. 13. Castor "The Lesser Ricinus"

Crispin van de Passe, *Hortus Floridus*, 1614

In mid-eighteenth-century North Carolina caraway was sown "in the meadows" of the new Moravian settlement. A ready source of caraway was desirable for remedies and for seasoning.[23]

> A little caraway is a welcome addition to either type of cabbage.
>
> —Von Rumohr, *Essence of Cookery*, 152

> [Along with chopped sage] carraway-seeds are used to flavour cheese.
>
> —Dods, *Cook and Housewife's Manual*, 505

Castor, Caster-Oil Nut, Palma Christi.

Ricinus communis

Mid-eighteenth-century Moravian settlers newly arrived in piedmont North Carolina received "a package of Palma Christi seeds." And Doddridge reported from western Virginia and Pennsylvania that "palma christi, the plant which furnishes the beans of which the castor oil is made . . . flourishes."[24]

> [Castor] Oil, which is essential to the preservation of health in every family, is made from the seed of a plant called Palma Christi, which is a native of most countries lying within the tropics, and will grow and flourish in all temperate latitudes.
>
> —Gunn, *Domestic Medicine*, 406

> These are violent Purgers, both by Vomit and Stool; and therefore hardly ever met with in Practice, unless amongst Empirics, and Persons of no Credit.
>
> —Quincy, *English Dispensatory*, 104

Caution: The bright red castor seeds are the source of the highly toxic protein ricin. If eaten, these seeds can cause a violent and potentially lethal gastrointestinal action.

Catmint, Nep. *Nepeta racemosa*; *Nepeta cataria*

Catmint may seem a bit out of place to a twenty-first-century herb gardener, but both John Lawson's 1709 journal and the Moravian diaries document early Americans' interest in growing nep in their kitchen gardens.

> [Catmint] is of special Use for the Windiness of the Stomach and Belly . . . for Colds, Coughs, and Shortness of Breath.
>
> —Culpeper, *English Physician*, 230

> [Nep] is a present help for them . . . that are very much bruised, if the juice be given with wine or mede [mead, a fermented beverage made from honey]. It is used in baths and decoctions for women to sit over, to bring downe their sicknesse, and to make them fruitfull.
>
> —Gerard, *Herball*, 683

> Catnip, particularly the blossoms, made into tea, is good to prevent a threatened fever.
>
> —Child, *American Frugal Housewife*, 37

Today's best-known name for this herb is curious—"catnip" being the combination of two ancestral common names, "catmint" and "nep." Cats' fondness for this herb has not changed through the centuries. They may be less attracted to established plants than to cut sprigs or newly set plants that seem to offer an invitation to "wallow or tumble in it."

Cats are very much delighted herewith; for the smel [*sic*] of it is so pleasant to them, that they rub themselves upon it, and wallow or tumble in it, and also feed on the branches and leaves very greedily.

—Gerard, *Herball*, 683

Chervil. *Anthriscus cerefolium*

Christian Reuter, a surveyor visiting the Moravian settlements in North Carolina, wrote that chervil "does not grow very well," although kerbel (chervil) was otherwise noted as established in their "upland garden."

The tender *Cimae* [blossom], and Tops with other Herbs, are never to be wanting in our Sallets, (as long as they may be had) being exceeding wholsome and chearing the Spirits: The Roots are also boil'd and eaten Cold; much commended for Aged Persons: this (as likewise Spinach) is us'd in Tarts, and serves alone for divers Sauces.

—Evelyn, *Aceteria*, 146

Much cultivated by the French and Dutch, who are so fond of it, that they have hardly a soup or salad but the leaves of chervil make part of the composition; and it certainly is often found a more agreeable and mild addition to seasonings, than parsley which is so universally used by the English cooks.

—Phillips, *History of Cultivated Vegetables*, 1: 146

I do use to eat them with oile and vinegar, being first boiled, which is very good for old people that are dull and without courage: it rejoiceth and comforteth the heart, and increaseth their strength.

—Gerard, *Herball*, 1040

[Chervil] is by many, esteemed in sallads and soups.

—M'Mahon, *American Gardener's Calendar*, 317

Its leaves are a good deal like those of double parsley. They are used in salads.

—Cobbett, *American Gardener*, 117

[Celery and sorrel] are usually accompanied by parsley, chervil, purslane and tarragon in the stock pot. Any imaginative cook will know how to vary quantities and proportions according to circumstances.

—Von Rumohr, *Essence of Cookery*, 161

Chervil Soup

Take a handful of chervil and half as much sorrel. Chop finely with a little parsley. Sauté this in a small piece of butter and add a spoonful of flour. When it completely cooks together, fill the pot with good meat broth. Also add salt and mace. Prior to serving, beat 2, 3 or 4 egg yolks together with a spoonful of sweet cream. [Add eggs and cream carefully to soup.] This is poured over sliced milk bread or sliced rolls.

—*Instructions of All Kinds of Cookery*, in Roberts, *The Raised Hearth*, 26

Chives, Cives. *Allium schoenoprasum*

Eighteenth-century British cookery books reveal a range of interesting and often complex seasoning blends. We expected to discover exotic and unusual combinations of herbs in French receipts but were surprised to find the overwhelming majority of French sauces and broths seasoned simply with parsley and chives or other green onions, along with the ubiquitous butter, wine or vinegar, and perhaps a little nutmeg or mace. The French allowed basic flavors of vegetables and meats to shine through. Both groups enjoyed chives despite ancient warnings that chives might be "hurtful to the eyes and braine."

Cives . . . are useful . . . both in sallads and culinary purposes; they are propagated by slipping the roots. . . . In slipping or parting the above roots, observe to preserve eight, ten, or more, of the small bulbs together in a cluster, and in that manner to plant them.

—M'Mahon, *American Gardener's Calendar*, 190

Cives, a little sort of onion, which is perennial. The greens only are used. A small quantity is sufficient for a garden.

—Cobbett, *American Gardener*, 117

The leaves... are often eaten with bread and butter by country people.

—Phillips, *History of Cultivated Vegetables*, 2: 33

A good pot-herbe . . . but evill for the eyes.

—W. Lawson, *Country House-wives Garden*, 79

Cives . . . ingender hot and grosse vapours, and are hurtful to the eyes and braine. They cause troublesome dreames, and worke all the effects that the Leeke doth.

—Gerard, *Herball*, 177

See also **Onion**.

Clary, Clary Sage, "or more properly" *Clear Eye.*

Salvia sclarea

Although we found clary in lists of Carolina gardens, we are left wondering how it was being used. Were early Americans following traditions of their homelands and harvesting clary to flavor cordials, omelets, or fritters? Or did they plant clary for its ancient reputation as "a gallant remedy for dimness of sight?"[25]

The Shops have a Spirit . . . an Infusion of the Herb [clary] in Spirit of Wine, and is very pleasant and a good Cordial.

—Quincy, *English Dispensatory*, 95

The Seed or Leaves taken in Wine provoketh to Venery [pursuit of sexual pleasure]. . . . It is of much use both for Men and Women that have weak Backs, and helpeth to strengthen the Reins. . . . This, they think will cure their Disease whereas . . . their backs are as much the better, as though they had pissed in their shoes.

—Culpeper, *English Physician, 90*

Clary when tender . . . in Omlets, made up with Cream, fried in sweet Butter, are eaten with Sugar, Juice of Orange, or Limon.

—Evelyn, *Aceteria*, 146

To make Clary Cake

Beat six Eggs very well with Salt and Nutmeg; spread a Handful of Clary, and mix with them; fry them, or put it in a Dish in the Dripping pan when Meat is roasting, and it will bake. You may make one the same Way with Chives and Parsley.

—Cleland, *Method of Cookery*, 121

To make Clarye Fritters

Take your Clarye-leaves, cut off the Stalk, dip them one by one in a Batter made with Milk and Flour, your Butter being hot, fry them quick. This is a pretty heartening Dish for a sick or weak Person; and Camfary-leaves [comfrey] done the same way.

—Glasse, *Art of Cookery* (1747), 82

The common name "clear eye" derives from several uses in remedies for the eyes. One might argue that the following treatment for cataracts is neither handsome nor safe as Culpeper asserted, although it certainly is interesting.

> The Seed is used to be put into the Eyes to clear them from Motes and such like things. . . . [Clary] is a gallant Remedy for Dimness of Sight, to take one of the Seeds of it, and put into the Eyes and there let it remain till it drops out of itself, the Pain will be nothing to speak on; it will cleanse the Eyes of all filthy and putrefied Matter, and . . . will take off a Film which covereth the Sight; a handsomer, safer, and easier Remedy . . . than to tear it off with a Needle.
>
> —Culpeper, *English Physician*, 90, 92

See also **Sage**.

Cleavers, Clivers, Goose-Grass, Ladies Bedstraw, Cheese-Rennet, Rennet-Wort.

Galium species

The smoother bedstraws or cleavers were relished among springtime greens "to cleanse the Blood," restoring health after sparse winter fare. Culpeper also described less appealing members of this group: "The common Cleavers, have diverse very rough square Stalks. . . . Stalks, Leaves, and Seeds, are so rough, that they will cleave to any thing that shall touch them," continuing, "It is familiarly taken in Broth to keep them lean and lank that are apt to grow fat." The Velcro-like texture of the common cleavers would indeed discourage overeating.

> The tender Winders [tendrils] . . . are us'd in Lenten Pottages [soups].
>
> —Evelyn, *Aceteria*, 147

Galium species also functioned (as did artichoke) as a plant source for rennet to "turn," or curdle, milk. Culpeper observed, "Besides the common Name [Ladies Bedstraw] . . . it is called Cheese-Rennet."[26]

> [Ladies Bedstraw] used for Rennet [or in addition to rennet from a calf's stomach] to make cheese . . . to turne their milke, that the Cheese which they make . . . might be the sweeter and more pleasant in taste, and also more wholesome.
>
> —Gerard, *Herball*, 1128

> To make an artificial Rennet, which will do very well for making of Cheese; and that is to boil the Cliver, or as some call it Goose-grass, or others Rennet-Wort, in Water, and you may add some Tops of Sweet Bryar; about a Spoonful of which Decoction . . . will turn a Pail-full of Milk, of about five Gallons . . . for the Improvement of the Cheeses, in giving them rich Flavours, it is adviseable to insert . . . either some of such Sweet Herbs as we like, or such Spices as we most covet the taste of.
>
> —Bradley, *Country Housewife*, 1: 76

> Ladies Bed-straw . . . is by some commended for consolidating and stopping Haemorrhages, yet it is not now much in use.
>
> —Quincy, *English Dispensatory*, 95

> [An infusion of cleavers] drank freely, is good for gravelly complants, and all obstructions of the urine.
>
> —Howard, *Botanical Medicine*, 250

Comfrey. *Symphytum officinale*

The 1633 edition of Gerard's herbal characterized this herb: "Comfrey joyeth in watery ditches, in fat and fruitfull meadows, they grow all in my Garden." The comfrey root was considered so effective for healing

FIG. 14. Comfrey

John Gerard, *The Herball*, 1633

fresh wounds "that it will sodder or glew together meate that is chopt in peeces seething in a pot, and make it one lumpe." We will leave it to the curious reader to test that claim.[27]

Although journals of John Lawson and Moravian settlers noted that comfrey grew in North Carolina, by the end of the eighteenth century professional medical practitioners were downplaying its efficacy: "Many ridiculous histories of the consolidating virtues of this plant are related by authors. At present [1791] it is so little employed in practice in Britain, as to have no place in our pharmacopoeias." American sources, however, continued to endorse comfrey for healing wounds into the nineteenth century.[28]

> *To draw out a Thorn*
> Take the roots of comfry, and bruise them in a mortar with a little boars-grease, and use this as a plaister.
> —Smith, *Compleat Housewife*, 326

> For wounded tendons: Boil comfrey roots to a thick mucilage or jelly, and apply this as a poultice.
> —Wesley, *Primitive Remedies*, 133

> [This] plant is useful to heal wounds.
> —Gardiner and Hepburn, *The American Gardener*, 35

Cucumber, or Cowcumber. *Cucumis sativus*
Wilde Cucumber. *Cucumis anguria*

John Lawson recorded a variety of "Cucumbers long, short, and prickly" in the Carolinas in 1709. The Virginia *Treatise on Gardening* advised: "If Cucumbers are stuck [trained to climb a stick] as you do pease, they will run to great height, and will bear till the frosts destroy them." Indeed, our experiments with wild cucumber, or burr gherkin, a small prickly cucurbit originating in Africa, produced bountiful fruits from midsummer into November in piedmont North Carolina. A tasty, hardy addition to the garden, this prolific plant provides a fresh vegetable when others are spent for the season, an important attribute in a frontier garden.[29]

> [Cucumber] is esteemed in its season the most refreshing and delecate of all vegetables.
> —*A Treatise on Gardening*, 19

> Take the Cucumbers and bruise them well, and distil the Water from them. . . . The Face being washed with the same Water, cureth the reddest Face that is: it is also good for Sun-burning, Freckles, and Morphew [blemishes].
> —Culpeper, *English Physician*, 110

> In many countries they use to eate Cowcumbers as wee doe Apples or Peares, paring and giving slices of them, as we would to our friends.
> —Parkinson, *Paradisi*, 524

Note: The long-held Galenic or humoral theory taught that good health was achieved through balancing the four humors (black bile, yellow bile, blood, and phlegm). Likewise, maintaining good health was best ensured by partaking of balanced foods. Certain fruits, herbs, and spices were believed to have such virtues as *hot* or *cold* and *dry* or *moist* and therefore were therapeutic for attaining balance. For example, cucumbers, thought cold and moist, could be balanced by adding the hot property of capsicum peppers.

FIG. 15. Common Cucumber, or Cowcumber "according to the Pronunciation of the Vulgar"

John Gerard, *The Herball*, 1633

FIG. 16. Wilde Cucumber

John Gerard, *The Herball*, 1636

FIG. 17. **"A trick of Robert Wilson and Mrs Kelly," 1812**

"She Saying robert what is good to distroy the bugs at the cucumber veins. He Said take turpentine and Sprinkle them. fillis came with turpentine and done so—distroyat bugs and vines."

Lewis Miller (1796–1882), from *Sketches and Chronicles*, 1966

> They are very cooling and viscid; and therefore apt to dispose the Blood to putrid Fermentations. . . . They do well therefore, who dress them with a good quantity of Vinegar and Pepper.
>
> —Quincy, *English Dispensatory*, 218

> Some use to cast a little salt on their sliced Cowcumbers, and let them stand halfe an houre or more in a dish, and then poure away the water that commeth from them by the salt, and after put vinegar, oyle, &c. thereon, as every one liketh: this is done, to take away the overmuch waterishness and coldnesse of the Cowcumbers.
>
> —Parkinson, *Paradisi*, 524

> Cucumber, tho' very cold and moist, the most approved Sallet alone, or in Composition . . . if rightly prepar'd; that is, by rectifying the vulgar Mistake of altogether extracting the Juice.
>
> —Evelyn, *Acetaria*, 147

Cucumbers were commonly pickled in a vinegar solution strongly flavored with hot spices such as pepper, ginger, cloves, nutmeg, and/or mace to counteract the cucumber's supposedly unwholesome coldness. Some authors included dill or fennel, but not sugar. Cucumbers were also cooked: stewed, fried, even forced (stuffed). English cookery books offered variations such as this one:

To Ragoo Cucumbers
Take two Cucumbers, two Onions, slice them, and fry them in a little Butter; then drain them in a Sieve, put them into a Sauce-pan, add six Spoonfuls of Gravy [broth], two of White Wine, a Blade of Mace; let them stew for five or six Minutes; then take a Piece of Butter as big as a Wallnut rolled in Flour; shake them altogether; and when it is thick, dish them up.
—Glasse, *Art of Cookery* (1747), 56

And a century later in South Carolina:

Stewed Cucumbers
Cut your cucumbers into thick slices, add some chopped onions, if liked, and some salt; let them simmer over a slow fire, till done enough; then pour off a large portion of the liquor, and add a little vinegar, pepper, butter and flour; let them stew a few minutes longer, and serve them up with the sauce.
—Rutledge, *Carolina Housewife*, 100

Garden predators were a constant problem. A few holly branches spread around a newly seeded cucumber bed were said to discourage chickens or cats from digging there, and onion was reputed to be a good insect repellent.

To Save Cucumbers from bugs, Salem Almanack advertisement, 1828
Set out an Onion or set up an Onion Stalk . . . in each hill and the streaked bugs will keep away.
—Lenoir, Medical memorandum

Dianthus.

See **Gillyflower**.

Dill. *Anethum graveolens*

We were not surprised to find dill in receipts for cucumber pickles written three centuries ago, "giving unto the cold fruit a pretty spicie taste." Compared with related seeds (fennel, caraway, anise), dill's "taste is moderately warm and pungent." Each of these seeds is "recommended as a carmenative in flatulent cholics."[30]

The leaves of Dill are much used in some places with Fish, as they doe Fenell; but because it is so strong, many doe refuse it. It is also put among pickled Cowcumbers, wherewith it doth very well agree, giving unto the cold fruit a pretty spicie taste or rellish. [Dill] being stronger than Fenell, is of the more force to expell winde in the body. Some use to eate the seed to stay the Hickocke [hiccup].
—Parkinson, *Paradisi*, 494

It stayeth the Hiccough, being boiled in wine, and but smelled unto, being tied in a Cloth.
—Culpeper, *English Physician*, 113

This plant is extreemly valuable as an ingredient in pickles, to which it gives a most exquisite flavour; the seeds when ripe are frequently used for that purpose, but it is the more general practise when they are formed, and not yet perfect, to cut off the umbels or heads, and then use them as above.
—M'Mahon, *American Gardener's Calendar*, 199

In North Germany there is one garden herb which traditionally accompanies broad beans and is therefore known as "bean herb" [*Bohnenkraut*]. This is dill, which is otherwise almost exclusively used in Pickling cucumbers.
—Von Rumohr, *Essence of Cookery*, 163

Fennel. *Foeniculum vulgare*
John Lawson found both "sweet and the common Sort" of fennel as he traveled through the Carolinas in 1709. The North Carolina Moravians recorded fennel in "the garden and wherever it is planted."[31]

> [Fennel] is exceeding good for the eyes. . . . [I]t is used in dressing Hives for swarmes [of bees]; a very good Pot-herbe, or for Sallets.
> —W. Lawson, *Country House-wives Garden*, 80

> Both Leaves, Seeds, and Roots thereof, are much used in Drink or Broth, to make People more spare and lean that are too fat.
> —Culpeper, *English Physician*, 133

> One good old Fashion is not yet left off, to boil Fennel with Fish; for it consumes that phlegmatick Humour, which Fish most plentifully afford and annoy the Body . . . though few that use it know why or wherefore they do it.
> —Culpeper, *English Physician*, 132

> In winter, the seed, bruised, give fish the same flavour as the leaves do in summer; and, to my taste, butter, seasoned with fennel, is better than any of the fish sauces, bought at the shops.
> —Cobbett, *American Gardener*, 124

> *Cure for the Colick*
> In distilling brandy, when it is run off, instead of running off the backings in the usual way, uncap your Still and put in near a peck of Tops of Fennel and run off one gallon and take a wine glassful night and morning.
> —Lenoir, Medical memorandum

Fennel, dill, anise, and caraway seeds shared a reputation as carmenatives, relieving flatulence and generally calming all sorts of gastrointestinal distress, such as "the heart burning." Each adds its own distinct savor to foods.

> *The Heart Burning*
> Chew fennel or parsley, and swallow your spittle.
> —Wesley, *Primitive Remedies*, 79

Feverfew, Featherfew. *Tanacetum parthenium* (syn. *Matricaria parthenium*, *Chrysanthemum parthenium*)
This inconspicuous relative of the asters was a well-known resident in early gardens. Although the name implies remedy for fever, the medical virtues of this plant range widely. Feverfew was administered, in beverages or external application, to "procureth women's sickness with speed," "to all hot inflamations or hot swellings," even to "draweth away fleghm and melancholy." Southern manuscripts included feverfew in salves, ointments, and poultices. By 1791, however, the Edinburgh *New Dispensatory* noted it was "now so little employed in Britain."[32]

> Fether-few shakes seed. Good against a shaking Fever [perhaps malaria], taken in a posset drink fasting.
> —W. Lawson, *Country House-wives Garden*, 80

> It is very effectual for all Pains in the Head coming from a cold Cause, the Herb being bruised and applied to the Crown of the Head. . . . The Decoction thereof drank warm.
> —Culpeper, *English Physician*, 131–32

FIG. 18. Fennel

Members of the Umbelliferae family such as fennel, carrot, and parsley are favored foods of black swallowtail butterfly larvae.

Maria Sibylla Merian, *Erucarum Ortus*, 1718

It is profitably applied to Saint Anthonies fire, to all hot inflamations, and hot swellings, if both leaves and floures be laid unto.
—Gerard, *Herball*, 653

An approved receipt for Worms
An handful of Rue the same quantity of Featherfew bruised fine nine large Live Earth Worms all sewed up in a fine bag & hung round the person against the Navel this is to be done either at full or new Moon for twenty four Hours—if the Person has Worms the Earth Worms will be totally Evaporated In that Case a Dose of Senna or Rhubarb should be given when the Bag is taken off.
—Unknown author, Herbal and household book

Garlic. *Allium sativum*

Although garlic was uncommon in eighteenth-century cookery, some enjoyed its strong flavor, especially in pickles. By the early nineteenth century we find Mary Randolph exceptional in her culinary use of garlic. Her *Virginia Housewife* called for garlic in several meat dishes (beef, veal, and mutton) as well as in pickles and sauces such as walnut and mushroom catsups and lemon pickle. A personal cookery manuscript from 1816 central North Carolina gives us one of the newly fashionable sauces of tomatoes, an early tomato catchup augmented with garlic (see figure 59).[33]

Almost all nations except the English, the Americans, and the French, make great and constant use of Garlick; and even the French use it, frequently, to an extent that it would drive us from the table.
—Cobbett, *American Gardener*, 124

Garlicks, tho' used by the French, are better adapted to the uses of medicine than cookery.
—Simmons, *American Cookery*, 37

Garlic was more commonly employed in domestic medicine for a range of complaints. In the Carolina backcountry Rachel Allen echoed John Wesley's popular treatment for earache and hearing loss: insert a clove of garlic dipped in honey into the ear. Wesley further recommended a clove of garlic on a painful tooth. William Lenoir noted a general therapy of garlic on the soles of the feet for sore throat, as did Wesley for hoarseness. In the South Carolina lowcountry, Harriott Horry suggested similar treatment for croup. Wesley tried an application of garlic salve for back pain similar to Horry's garlic pill for rheumatism.[34]

Very strong smell and taste . . . but exceeding wholesome withall for them that can take it. . . . [Garlic is] called in divers Countries, the poore mans Treakle, that is, a remedy for all diseases.
—Parkinson, *Paradisi*, 514

The most commodious form for taking garlick, a medicine to most people not a little unpleasant, is that of a bolus or pill.
—*Edinburgh Dispensatory*, 121

Cure a Cough
Put Garlick in Whiskey and drink frequently of it.
—Lenoir, Medical memorandum

Hooping Cough [from] Mrs Myddleton
The External is three Cloves of Garlic bruised very small then well mixed with one ounce of new unsalted Butter applied . . . to the Soles of the feet.
—Unknown author, Herbal and household book

An Old Stubborn Pain in the Back
Apply garlic and hog's lard to the feet. (Tried)
—Wesley, *Primitive Remedies*, 93

See also **Onion**.

Gillyflower, Clove Gilliflower, July-Flower, Pinks, Clove Pinks. *Dianthus caryophyllus* or perhaps other *Dianthus* species

According to *The Country House-wives Garden*, the clove pink was highly regarded in 1660 England: "of all flowers (save the Damask Rose) they are the most pleasant to sight and smell. . . . Their use is much in ornament, and comforting the spirits, by the sense of smelling." Their spicy fragrance was also captured to flavor conserves and syrups. In North Carolina a hundred years later, Moravian chroniclers noted clove pinks "have been brought here, and . . . do well."[35]

> There are many kinds of Gillofloures, Pinkes, and the like, which differ very little in their roots, leaves, seeds, or manner of growing, though much in the colour, shape, and magnitude of their floures. . . .
>
> The conserve made of the floures of the Clove Gillofloure and sugar is exceedingly cordial, and wonderfully above measure doth comfort the heart, being eaten now and then. . . .
>
> [Clove Gillyflowers] are not used in Physicke, but esteemed for their use in Garlands and Nosegaies. They are good to put into Vinegar, to give it a pleasant taste and gallant color.
>
> —Gerard, *Herball*, 589, 590, 597

FIG. 19. Gillyflowers, Pinks

John Parkinson, *Paradisi in Sole*, 1629

> At present the flowers are chiefly valued for their pleasant flavour, which is entirely lost even by light coction; hence . . . the syrup, which is the only officinal preparation of them, to be made by infusion.
>
> —*Edinburgh Dispensatory*, 164

> *To make Syrup of Clove-Gilliflowers*
> Clip your Gilliflowers, and sprinkle them with fair Water, put them into an earthen Pot, stop it up very close, and set it in a Kettle of Water, and let it boil for two Hours; then strain out the Juice, put a Pound and half of Sugar to a Pint of juice, put it into a Skillet, set it on the Fire, keeping it stirring till the Sugar is all melted; but let it not boil, then set it by to cool, and put it into Bottles.
>
> —Glasse, *Art of Cookery* (1747), 153.

Similar to Pre-1744 Virginia cookbook

Gourd. *Cucurbita* or *Lagenaria* species and others

Early references to these fruits are at times confusing, especially because winter squashes were sometimes called gourds, and gourds might be called calabashes. The gourds of our region grow on vines like their relatives the squashes and pumpkins. True calabashes are gourdlike fruits of a tropical tree.

> The Planters put Gourds on standing Poles, on purpose for these Fowl [Martins] to build in, because they are a very Warlike Bird, and beat the Crows from the Plantations.
>
> —J. Lawson, *New Voyage*, 149

> They are kept for the rinds wherein they put Turpentine, Oyle, Hony, and also serve them for pales to fetch water in.
>
> —Gerard, *Herball*, 923

> Callabasch, a kind of Gourd. All sorts of vessels can be made from them; the hunters use them for powder flasks.
>
> —Fries, *Records of the Moravians*, 2: 573

> In order to make vessels of [gourds], they are first dried well; the seeds, together with the pulpy and spongy matter in which they lie, are afterwards taken out and thrown away; the shells are scraped very clean within, and then great spoons or ladles, funnels, bowls, dishes, and the like may be made of them: they are particularly fit for keeping seeds of plants in, which are to be sent over sea, for they keep their power of vegetating much longer, if they be put in calabashes [gourds], than by any other means. Some people scrape the outside of the calabashes before they are opened, dry them afterwards, and then clean them within; this makes them as hard as bones; they are sometimes washed, so that they always keep their white colour.
>
> —Kalm, *Travels*, 1: 183

> [Gourd vines provide] a covering for arbours and walking places.
>
> —Gerard, *Herball*, 925

Hops. *Humulus lupulus*

The *Records of the Moravians in North Carolina* reported, "Hops grow in sunny Bottoms; our cultivated ones are taken from them." Miller's *Gardeners Kalendar*, Tusser's *Good Husbandry*, and other sources directed the gardener to "pole Hops" in springtime. Hops vines allowed to grow up simple poles could be harvested by cutting the vine a few inches above ground level and carrying the pole, with vines still attached, to a spot more convenient for those picking the blossoms—perhaps under a shade tree.[36]

FIG. 20. Gourd "Purple Martin (Calabash)"

John James Audubon, *Birds of America*, vol. 1, 1849

Hops blow [bloom] in July, bell [form seed heads] in August, and are ripe by the first of September. When they begin to change colour, they are easily pulled to pieces, and their seeds look brown within them, they are ripe, and are then to be gathered as soon as possible.
—*Valuable Secrets*, 205

Sept. 23, 1773: The sisters have finished picking hops; there has been a good crop.
—Fries, *Records of the Moravians*, 2: 780

The floures are used to season Beere or Ale with, and too many do cause bitternesse thereof, and are ill for the head. . . . The manifold vertues of Hops do manifestly argue the wholsomenesse of beere above ale; for the hops rather make it a physicall drinke to keepe the body in health, than an ordinary drinke for the quenching of our thirst.
—Gerard, *Herball*, 885

The bright hops of last year's growth are best, the older the worser. Hops ripe, when gather'd free of the damage of insects and rains, and if rightly kiln-dried, will (by rubbing them between the finger and thumb) feel oily, yield a delicate scent, and taste thoroughly bitter.
—Ellis, *Country Housewife's Family Companion*, 459

The floures make bread light . . . if the meale be tempered with liquor wherein they have been boyled.
—Gerard, *Herball*, 885

Yeast
Take 5 common sized Irish potatoes and boil them in about 4 quarts of Water till they are soft enough to mash. Then add a handful of hops and let it boil up, then take it off the fire let it cool a little and strain it, and put a large spoon-ful of Brown sugar in it. Bottle it and keep it warm till it works then put it in a cool place but don't cork it. . . .

FIG. 21. Hops Picking (c. 1804)

W. H. Pyne, *Picturesque Views of Rural Occupations*, 1808

When this yeast is good one pint is sufficient for 2 loaves of Bread.

—Horry, *Colonial Plantation Cookbook*, 95

Miller's popular English *Gardeners Kalendar* and others reported that "hop-tops are often eaten when there is a scarcity of other greens." On the other hand, the hops blooms were thought "rather Medicinal, than fit for Sallet."[37]

In the months of March and April, while the buds are tender, the country people dress them as asparagus; they are an agreeable vegetable, and esteemed good to purify the blood in the scurvy, and most cutaneous [skin] diseases.

—Phillips, *History of Cultivated Vegetables*, 1: 248

To range the Hop amongst Vegetables may appear odd; but, it is a garden plant in America, and does give . . . a very good dish for the table. . . . [When] three inches long, you may cut, and eat as asparagus, cook [the hop shoots] in the same manner, and you will find them a very delightful vegetable.

—Cobbett, *American Gardener*, 124–25

Hops Soup

Take chervil, parsley, sorrel and onion. Chop small and sauté a little in butter. Pour meat broth over it. Meanwhile, cook hops [tender buds or shoots] or asparagus in salt water, but not too soft. Cut this into small pieces and add to the broth. Stir in a few egg yolks before serving. Serve over toasted slices.

—*Instructions of All Kinds of Cookery*, in Roberts, *The Raised Hearth*, 33

Hops blooms were looked to for pain relief. An eighteenth-century household book from an unknown South Carolinian included the following instruction for treating sore throat: "Take a handful of hops, steep in spirits. . . . Let the hops be applied like a poultice." In 1814 William Lenoir copied similar advice from the *Raleigh Star* into his own medical memorandum book. A nineteenth-century medical guide suggested a related remedy for pain:[38]

Hops are also a very valuable external application for pains, especially of the spasmodic kind. For this purpose they may be put into a small bag, thoroughly moistened with hot vinegar, and applied to the painful part.

—Howard, *Botanical Medicine*, 255

The odour of hops hung in a bed has been said to induce sleep after opium had failed.

—*Edinburgh Dispensatory*, 224

Horehound. *Marrubium vulgare*

Our traditional use of horehound for cough relief has been recognized for centuries, remaining essentially unchanged from the ancient herbals well into the eighteenth and nineteenth centuries. Horehound syrups and lozenges remain a popular twenty-first-century home remedy today. Note the earlier advice, as shown in quotations from Culpeper, Thomson, and Wesley below:

There is a Syrup made of Horehound to be had at the Apothecaries, very good for old Coughs.

—Culpeper, *English Physician*, 174

Lozenges made of the juice of this herb and sugar, are esteemed good for colds that affect the chest.

—Phillips, *History of Cultivated Vegetables*, 1: 250–51

For an inveterate cough . . . take a spoonful of syrup of horehound morning and evening.

—Wesley, *Primitive Remedies*, 52

It is rarely prescribed by physicians [by 1802]. . . . An infusion of half a handful of the fresh leaves . . . is usually practiced by the common people, with whom it is still a favourite remedy in coughs and asthmas.

—Thomson, *Family Physician*, 397

Horehound's long-held reputation against poisons was also alive and well in the early American South. A combination of horehound and plantain juices for treating poisoning and the bite of snake, spider, or mad dog was developed and made famous by an enslaved African in South Carolina named Caesar. Widely known as Caesar's Cure, directions for its preparation were published in newspapers and almanacs, entered into personal copybooks, and shared person to person. In recognition of this important discovery, the South Carolina General Asssembly rewarded Caesar by granting him freedom and an annual sum of one hundred pounds for the duration of his life.[39]

Syrup made of the [horehound] green fresh leaves and sugar. . . . Is likewise good for them that have drunke poyson, or that have beene bitten of Serpents.

—Gerard, *Herball*, 694

Roots and leaves are used as an antidote for poison and Rattlesnake bites, and are called Caesar's Cure.

—Fries, *Records of the Moravians*, 2: 574

This plain and simple remedy may be readily obtained, and easily administered by the most ignorant, as both horehound and plantain grow spontaneously near almost every house.

—Smyth, *A Tour in the United States*, 109

Negroe Caesars cure for the bite of a Rattlesnake
Take the roots of plantain or hoarhound (in Summer and branches together) in a Sufficient quantity, bruise them in a morter and Squeeze out the Juice of which as soon as possible take one Spoonfull; if he is Swelled you must force it down his throat.

—Lenoir, Medical memorandum

Horseradish. *Armoracia rusticana* (syn. *Cochlearia armoracia*)

Horseradish was cultivated in gardens for medicinal and culinary uses. Gerard's *Herball* noted an early German fondness for horseradish with seafoods that continues today in our standard seafood sauces. In England, wild horseradish was gathered and combined with mustard seed, creating a doubly spiced condiment, Tewkesbury mustard, named for the market town where it originated. Today's popularity of horseradish with beef can likewise be traced from England to America, as seen below.

Horse Radish stamped with a little vineger put thereto, is commonly used among the Germans for sauce to eate fish with, and such like meates, as we [English] doe mustard; but this kinde of sauce doth heate the stomacke better, and causeth better digestion than mustard.

—Gerard, *Herball*, 242

Get young, tender roots of horseradish; wash them clean, and scrape them remarkably fine with a sharp knife. Put it in a jar, with enough plain cold vinegar to cover it well; add no seasonings. Cover the jar securely with a piece of leather. Horseradish thus prepared will be found an excellent condiment for fresh meats.

—Bryan, *Kentucky Housewife*, 184

FIG. 22. Horse Radish

John Gerard, *The Herball*, 1633

[Horseradish] promotes digestion, and therefore is properly employed as a condiment with our animal food.
—W. Barton, *Cullen's Treatise*, 2: 121

Horseradish-wine . . . is recommended for gouty habits.
—Dods, *Cook and Housewife's Manual*, 470

[Horseradish's] great activity and warmth also make it good in all such nervous cases as . . . induce heaviness of the senses, or inaptitude to motion.
—Phillips, *History of Cultivated Vegetables*, 1: 257

Though [horseradish is] a very valuable and wholesome article of diet, it is a most pernicious weed.
—Cobbett, *American Gardener*, 127

This plant is cultivated by cuttings of the root, either cut from the top an inch or two long, or some old roots cut into pieces of that length, or by small offsets that arise from the sides of the main root, retaining the crowns or top shoots on as many as possible.
—M'Mahon, *American Gardener's Calendar*, 201

When the [horseradish] is planted, the ground may be sown with spinage or radishes, which will come off time enough to give the horseradish full time to grow. . . . In the autumn after planting, or about michaelmas [September 29], you may begin to take up some of the roots, if wanted for use: but it will be adviseable to let the principal part remain to have another summer's growth, when they will be very fine and large.
—Squibb, *Gardener's Calendar*, 15

Houseleek. *Sempervivum* species
Stonecrop, Sengreen, Sedum. *Sedum* species

Today the houseleek may be best known by its descriptive name "hen and chicks" because it propagates by sending out new rosettes (offsets) from a central rosette of foliage. Houseleeks and stonecrops have been encouraged for centuries to grow on roofs and stone houses, and these succulents are among species chosen to cover environmentally approved "green roofs" today.

These two groups of common garden, house, and pot plants have a long reputation as soothing to burns and skin irritations. As Gerard noted, houseleek and sengreen "take away the fire in burnings and scalding." And Tennessean John Thompson's poultice for treating the redness accompanying St. Anthony's fire included houseleek and sorrel.[40]

It is customary, with us, among the common sort . . . to give the expressed juice of houseleek and sugar, in fevers, and hot diseases.
—Phillips, *History of Cultivated Vegetables*, 1: 264

Our ordinary Housleek is good for all inward Heats, as well as outward. . . . It easeth also the Headach, and distempered Heat of the Brain in Frenzies, or through want of Sleep, being applied to the Temples and Forehead.
—Culpeper, *English Physician*, 177

The Sting of a Wasp
Rub the part with the bruised leaves of house-leek, water-cresses or rue.
—Wesley, *Primitive Remedies*, 112

Cure for Deafness from Farmer Almanac 1834
Equal parts of the Juice house-leak, Brandy and Sweet Oil in a phial to be hung up exposed to the sun a month or more then droped in the [ear] of night and on wool to be kept in the ear, a shur remedy for deafness.
—Lenoir, Medical memorandum

FIG. 23. Lavender

John Gerard, *The Herball*, 1633

Cure for Corns
First shave the corn off as close as you can then take house leak and bind a leaf to each corn.
—Thompson, Commonplace book

Ladies Bedstraw.

See Cleavers.

Lavender. *Lavandula angustifolia*

Lavender falls into the group of Old World herbs originating from the Mediterranean region and other dry climes. *A Treatise on Gardening* tells us, "It has been found that [a poor gravelly] soil suits it best, will give it more aromatic smell." Lavender is documented in American frontier gardens in John Lawson's list of "pot-herbs and others of use" and in the medical garden of the North Carolina Moravian settlers.[41]

The enchanting scent of lavender still delights us. Its current popularity in aromatherapy can be traced back centuries, for "Griefs and Pains," as well as "Tremblings and Passions," not ignoring "Faintings and Swoonings":

> Lavender is of a special good Use for all the Griefs and Pains of the Head and Brain . . . as also the Tremblings and Passions of the Heart, and Faintings and Swoonings, not only being drank, but applied to the Temples, or Nostrils to be smelt unto.
> —Culpeper, *English Physician*, 188

> [Lavender is] whether externally applied or given internally, a powerful stimulant to the nervous system.
> —W. Barton, *Cullen's Treatise*, 2: 105

> The flowers should be gather'd and dry'd in a shady Place, and preserv'd dry for Use.
> —P. Miller, *Gardeners Dictionary*

> This flowre is good for Bees, most comfortable for smelling, . . . and kept dry, is as strong after a year, as when it is gathered. The water of this is comfortable.
> —W. Lawson, *Country House-wives Garden*, 81

> It takes its Name of *Lavendo* [meaning] Washing; because it was used to be thrown into Baths, for the Fragrancy of the Scent; or . . . to give a Fragrancy to Linen, and because it is very good to wash the Face with, and give it both Beauty and a grateful Scent.
> —P. Miller, *Gardeners Dictionary*

Lavender Brandy [primarily a perfume or room freshener]
Put a quart of lavender flowers in a pitcher; fill it up with white brandy, and let it steep till next day; then squeeze them out, fill up your pitcher with fresh flowers, and repeat the process till you have a strong infusion of the flowers: then strain and cork it up in a bottle.
—Bryan, *Kentucky Housewife*, 435

Lettuces. *Lactuca sativa*

An array of lettuces was grown throughout this period. At the dawn of the seventeenth century Gerard listed eight; and in 1709 John Lawson found several lettuces growing in the Carolinas: "The Sallads are the Lettice, Curl'd, Red, Cabbage and Savoy. The Spinage round and prickly." Even dandelion, lambsquarter, violet leaves, and other wild greens were counted among the lettuces. Phillips in his 1822 *History of Cultivated Vegetables* professed he had "always found it desirable to have some dandelion plants taken from the pastures or road sides, and planted in our garden to blanch for the spring, as it is then an agreeable herb to mix with other salads, and may be procured when lettuce or endive are not easily obtained." Today we find it unusual that lettuces were not only "usuall in Sallets" but also "in the pot." Lettuce was combined with mint or sorrel in pea soups and in other springtime soup receipts. See "Potage à la Julienne" in **A Salmagundi of Methods** and the following German and French receipts:[42]

> The term, salad, is used to describe all the various types of lettuce which are used in the kitchen, although it actually denotes a particular method of preparation rather than a plant. . . . Like spinach, all of the lettuce varieties can be used as a vegetable. They can be included in a herbal spinach mixture and in herb soups. It is, however, more normal to serve them raw, dressed with oil, vinegar and salt like other salad vegetables.
>
> —Von Rumohr, *Essence of Cookery*, 161

> ***Potage à la Jardinière, or Gardener's Soup***
> This is like all other Spring soups [it begins with winter leftovers: carrots, turnips, onions, celery], only add leaves of sorrel and lettuce. . . . The cutting of the vegetables forms the only variety.
>
> —Ude, *French Cook*, 61

> The several sorts of garden lettuces are very wholesome, emolient, cooling salad herbs, easy of digestion, and somewhat loosening the belly. Most writers suppose that they have a narcotic quality.
>
> —*Edinburgh Dispensatory*, 217

> [Lettuce] inclines [one] to sleep: Upon which account it procures ease in Pains; both taken inwardly, and externally apply'd.
>
> —Quincy, *English Dispensatory*, 217

> This great article of the garden is milky, refreshing, and pleasanter to a majority of tastes than any other plant, the Asparagus hardly excepted. So necessary is it as the principal ingredient of a good salad that it is, both in France and England called "salad." . . . It is good in stews; good boiled with green-peas; and, even as a dish boiled as cabbage is, it is an excellent vegetable. Yet, I never saw a really fine lettuce in America [due to the difficulty of preserving plants in the winter and the "great heat" in late spring].
>
> —Cobbett, *American Gardener*, 128–29

The milky sap of some lettuces (for example, romaine) may have inspired the notion, in accordance with the Doctrine of Signatures, that lettuce "much increases Milk in the Breasts of such as give Suck."[43]

Mallows. *Malva* species; *Hibiscus* species
Marsh Mallow, Marshmallow. *Althaea officinalis*

It is rather difficult to determine which, among many possible mallows, is being referenced in each passage below. Regarding mallows in the Carolinas, "There are all kinds here" and of "several sorts." Mallows were appreciated for their beauty, of course, and also for their mucilaginous property. This "viscous or slimie quality" was thought to indicate usefulness in lubricating various passages within the body—particularly the urinary and gastrointestinal systems.[44]

FIG. 24. Marshmallows

Elizabeth Blackwell, *A Curious Herbal*, 1739

They are best when intermix'd with other Flowers of the same Growth, where they afford an agreeable Variety.
—P. Miller, *Gardeners Dictionary*

The wilde Mallow without controversie is fitter to be eaten, and more pleasant than those of the garden, except the French Mallow, which is generally holden the wholsommest.
—Gerard, *Herball*, 932

All sorts of Mallowes, by reason of their viscous or slimie quality, doe helpe . . . being used inwardly . . . ease the paines of the stone and gravell . . . being outwardly applyed, they mollifie hard tumors, and helpe to ease paines in divers parts of the body.
—Parkinson, *Paradisi*, 370

The Syrup also and Conserve made of the Flowers are very effectual.
—Culpeper, *English Physician*, 202

[Mallow leaves] were formerly in some esteem, in food, for loosening the belly . . . and a conserve was formerly prepared from the flowers.
—*Edinburgh Dispensatory*, 225

Costiveness [constipation]
Boil in a pint and a half of broth, half a handful of mallow leaves chopped, strain this and drink it before you eat anything else. Do this frequently if needful.
—Wesley, *Primitive Remedies*, 50

[For inflammations:] Marsh Mallows boiled in Milk applied as a Poultice.
—Unknown author, Herbal and household book

For the Vertigo, or Swimming in the Head
In a May morning about sun-rise snuf up the dew, daily, that is on mallow leaves.
—Wesley, *Primitive Remedies*, 123

Marigold, Marygold.
Calendula, Pot Marigold, Garden Marygold. *Calendula officinalis*
African Marigold. *Tagetes erecta*
French Marigold. *Tagetes patula*

The calendula, or pot marigold, was well known to English settlers in America, although the unrelated French and African marigolds (*Tagetes* species) with which we are today more familiar were also described in the early herbals. The common name "marigold" leaves us wondering just which species is cited in each reference below. The authors quoted here were likely considering *Calendula* as the marigold for use. Adding to the puzzle, a scrap of paper associated with an unidentified Virginia manuscript circa 1700 does list "french Marigoles" in an inscrutable list of plants.[45]

Glasse's *Art of Cookery* recommended adding marigold flowers along with other herbs to mutton broth or Scotch barley broth, following earlier traditions for marigolds in a variety of meat and poultry dishes.[46]

> The herbe and flowers are of great use with us among other pot-herbs and the flowers either greene or dryed, are often used in possets, broths and drinkes, as a comforter of the heart and spirits, and to expel any malignant or pestilential quality. . . . The Syrupe and Conserve made of the fresh flowers, are used for the same purposes to good effect.
> —Parkinson, *Paradisi*, 298

FIG. 25. Pot Marigold
(Bending-Stalked Marigold)
Calendula tragus

Unknown artist, 1798, *Curtis's Botanical Magazine*, vol. 12

FIG. 26. African and French Marigolds "Caryophyllus"

Basilius Besler, *Hortus Eystettensis*, 1613

The Plague (to prevent.) Eat marigold flowers, daily, as a salad, with oil and vinegar.
—Wesley, *Primitive Remedies*, 96

They are excellent in broths and soups and stews.
—Cobbett, *American Gardener*, 131

The double Marigold, being as bigge as a little Rose, is good for shew. They are a good Pot-herbe.
—W. Lawson, *Country House-wives Garden*, 81

To Pickle Marygold-Flowers
Though the Marygold Flower is of old date in Pottage; yet it continues still its stand among the useful things of the Kitchen . . . according to the custom of the Country, when the Country Folks make Broth. . . . And as the Marygold-Flowers are used in Porridge, I send the Receipt how to pickle them. Strip the Flower-leaves off, when you have gather'd the Flowers, at Noon, or in the Heat of the Day . . . pour the Salt and Water upon them; then shut them up close till you use them, and they will be of a fine Colour, and much fitter for Porridge than those that are dry'd.
—Bradley, *Country Housewife*, 2: 91

Calendula officianalis, or garden marygold. The dried flower-leaves are sometimes used as a yellow dye. Their expressed juice, boiled with alum, makes an excellent yellow paint.
—*1771 Encyclopaedia Britanica*, 1: 634

[When making a New-market cheese] colour it with marigolds.
—Smith, *Compleat Housewife*, 99–100

It was an old practice with dairy-women, to churn the petals of the marigold with their cream, to give their butter a yellow colour.
—Phillips, *History of Cultivated Vegetables*, 1: 332

To colour their Butter . . . they bruise the [Marigold] Flowers in a Mortar, then put them into a Rag, to squeeze out their Juice amongst the Butter, which, on being work'd in, will give it a fine yellow Colour and wholesome quality. . . . [L]ittle inferior to Saffron for many Uses.
—Ellis, *Country Housewife's Family Companion*, 322–23

The French and African marigolds with either single or double blossoms were of varying shades of yellow and orange, "some . . . beautifully variegated." The taller are the African marigolds, and the dwarf or creeping are the French. *Tagetes* thrive better than *Calendula* in the warmer, southern regions of America but were of debated value for culinary or medicinal use. Gerard's *Herball* posted a warning regarding the French or African marigold: "full of poison, and therefore not to be touched or smelled unto, much lesse used in meat or medicine." That early herbalist's warning may be a little overcautious, although sensitive individuals may suffer dermatitis after handling the common *Tagetes* marigold.[47]

Marjoram.

Sweet Marjoram, Knotted Marjoram, Summer Marjoram. *Origanum majorana* (syn. *Majorana hortensis*)

Pot Marjoram, Wild Marjoram, Winter Sweet Marjoram. *Origanum vulgare*

Both sweet or knotted marjoram (annual) and pot, wild, or winter marjoram (perennial, now commonly known as *oregano*) found a place in eighteenth-century gardens. In England and America marjoram was thought "excellent good" among sweet herbs for seasoning soups and boiled meats. The German author Von Rumohr bemoaned a surprising fall in popularity: "Nowadays (1822), marjoram and thyme are hardly used except in sausages, and then they are used in dried and powdered form. Employed sparingly, they can, however, give an excellent flavour to other chopped mixtures." Marjoram continued to hold a place in American kitchens and gardens. The pre-1744 cookery manuscript from Virginia suggested adding sweet marjoram and rosemary to the usual honey, lemon, and spice to make mead.[48]

> One sort is annual and one perennial. The former is called "summer" and the latter "winter." . . . Both ought to be preserved by drying.
>
> —Cobbett, *American Gardener*, 130

> *To make a Fricassy of Mushrooms*
>
> Make ready your Mushrooms [stew them and discard the liquor that comes from them] . . . put them into a Frying-pan with a piece of Sweet Butter little parsly, Thyme, Sweet Marjoram, a piece of Onion Shred very Small, a little Salt & fine beaten Pepper, fry them till they are enough, then having the . . . Lear [thickened sauce] ready, put it to the Mushrooms while they are in the pan toss them two or three times . . . and Serve them.
>
> [The lear for a mushroom fricassee may be made with mutton gravy (or other broth), the juice of a lemon (or claret wine) and orange, and a little grated nutmeg, stirred together with the yolks of two eggs per every cupful of liquid.]
>
> —Unidentified cookbook, circa 1700 Virginia, in Harbury, *Virginia's Cooking Dynasty*, 148, 152

> *To make Frost [forced] meat*
>
> Take Veal Lean beef or pork is best; take a pound of pork & Shred it very smal then take as much Sewett [suet] as meat & Shred it but not so smal, then take a Little Sprig of time [thyme] and Sweet Marjoram one nutmeg half an ounce of pepper with some Cloves & mace well beat and a handful of salt, then Roul it up into Small Balls.
>
> —"Jane Randolph her Cookery Book" (1743), in Harbury, *Virginia's Cooking Dynasty*, 312

> [Wild marjoram] has an agreeable aromatic smell, approaching to that of sweet marjoram, and a pungent taste much resembling thyme. . . . [Sweet marjoram] has long been cultivated in our gardens, and in frequent use for culinary purposes. The leaves and tops have a pleasant smell, and a moderately warm, aromatic, bitterish taste.
>
> —Thomson, *Family Physician*, 403, 404

> A few leaves of fresh marjoram will take the rancid flavour away from minced lamb and kid.
>
> —Von Rumohr, *Essence of Cookery*, 162

> The leaves are excellent good to be put into all odoriferous ointments, waters, pouders, broths, and meates.
>
> —Gerard, *Herball*, 665

> The sweete Marieromes are not only much used to please the outward senses in nosegayes, and in the windows of houses, as also in sweete pouders, sweete bags. And sweet washing waters, but are also of much use in Physicke.
>
> —Parkinson, *Paradisi*, 453

> [Marjoram] is principally celebrated in disorders of the head and nerves, and in the humoural asthmas and catarrhs [chest congestion] of old people.
>
> —*Edinburgh Dispensatory*, 225

It has the same . . . qualities with lavender [for head and nerves].

—W. Barton, *Cullen's Treatise*, 2: 106

Melissa.

See **Balm**.

Mints.

Peppermint. *Mentha × piperita*

Spear, Roman, Garden Mint (Curled or Variegated). *Mentha spicata*

Orange Mint. *Mentha citrata*

The extensive mid-eighteenth-century *Records of the Moravians in North Carolina* listed mint as growing in the medical garden and elsewhere. The Virginia *Treatise on Gardening* noted an interesting fact regarding this genus: "*Mentha* from *mens*, the mind, because it strengthens the mind." Mints entered into culinary, medical, and other uses. Mawe's 1773 *Every Man His Own Gardener* specified "Spear[mint], proper sort for all kitchen uses, Pepper[mint] for distilling, Orange[mint], principally for curiosity."[49]

[Peppermint] is commonly preferred to the other species of mint. . . . Useful in flatulent colics, hysterical affections, retchings, and other symptoms of indigestion.

—Thomson, *Family Physician*, 412

Peppermint is a warm stimulant to the stomach . . . holding a first rank in the list of medicines . . . which dispel, or scatter the wind from the stomach and bowels.

—Gunn, *Domestic Medicine*, 400

[In] a Fever, with violent Purging and Vomiting . . . be careful to live some Days upon light and innocent Fare, boiling Mint in every Thing you eat, and grating Nutmeg into all your Drink.

—Tennent, *Every Man His Own Doctor*, 32

[Spearmint] has a more agreeable flavour [than peppermint], and is therefore preferred for culinary uses, and more generally cultivated in our gardens.

—Thomson, *Family Physician*, 426

Mint Tea [also Sage or Balm]

Pick the leaves and stalks of spare [spear] mint, rinse them clean, put them in a pitcher, and pour boiling water on, cover it, and let it stand for a few minutes to infuse; then turn the tea from the leaves, sweeten it lightly with loaf sugar, and serve it warm. It is a sovereign efficacy in settling a sick stomach after taking an emetic. Sage tea is also good. Balm makes a very cooling tea; it is good to use in fevers as a constant drink.

—Bryan, *Kentucky Housewife*, 425

The leaves and tops of spear-mint are used in spring salads, as also in acid sauce with roasted lamb, &c. It is boiled with green peas, and generally used in pea soup on account of its carminative quality.

—Phillips, *History of Cultivated Vegetables*, 1: 344

[Spearmint] is used for the table, in many ways. The French snip a little into their salads, we boil a bunch amongst green peas, to which it gives a pleasant flavour; chopped up small, and put, along with sugar, into vinegar, we use it as sauce for roasted lamb; and a very pleasant sauce it is.

—Cobbett, *American Gardener*, 132

The following "Peas Porridge" receipt illustrates the eighteenth-century preference for mint as a seasoning for peas. Fresh green peas were widely enjoyed when available in springtime; during the rest of the year the homely dried pea provided a base for soups, puddings, and porridges. See another pea soup method under **Sorrel** that calls for sorrel, parsley, marigolds, and lettuce along with mint.

> ***Pease Porridge***
> Boil green pease quickly after they are shelled, and when they are boiled a little tender, we put some milk into the pot, and boil them about a quarter of an hour, then stir in some flower [flour], till the porridge is almost as thick as pap, and when all is boiled a little longer, with a bunch of mint, it is done enough. At last, some pepper and salt is put into each mess, and eaten generally without bread or butter, because in this way of preparing green pease porridge, it is thick enough without bread.
> —Ellis, *Country Housewife's Family Companion*, 292

> ***Green Pea Soup***
> Put a handful of mint chopped small, and a pint of young peas which must be boiled in the soup till tender; thicken it with a quarter of a pound of butter, and two spoonsful of flour.
> —M. Randolph, *Virginia Housewife*, 34

We find the early English taste for mint in peas continued into mid-nineteenth-century Carolina in *The Carolina Housewife*'s typical southern "Hopping John" of red peas, rice, and bacon. "Season with salt and pepper, and, if liked, a sprig of green mint."[50]

> The mint . . . will be in fine order for mint-sauce, &c. and either to use alone as a sallad, or to mix among other small herbs.
> —M'Mahon, *American Gardener's Calendar*, 11

Both *Virginia Housewife* and *Kentucky Housewife* gave recipes for mint sauces similar to the basic instructions from Scots author Margaret Dods. These sauces were particularly, but not exclusively, recommended to accompany lamb.

> ***Vinegaret for cold meat or fowl***
> Chop young mint, parsley, and eschalot together, and mix them up with salt, oil, and good vinegar.
> —Dods, *Cook and Housewife's Manual*, 240

> ***Lamb***
> [The lamb may be roasted or boiled] and served with mint sauce in a boat, chop the mint small and mix it with vinegar enough to make it liquid, sweeten it with sugar.
> —M. Randolph, *Virginia Housewife*, 56

> ***Warm Mint Sauce for Mutton and Lamb***
> Take equal portions of drawn butter and vinegar; put them in a sauce pan, with a very little pepper, and when it begins to boil, make it tolerably thick, with equal portions of mint and boiled yolks of eggs.
> —Bryan, *Kentucky Housewife*, 169

Mint was valued for attracting bees, as were rosemary, thyme, lavender, borage, and bugloss.

> It is a common practice to rub the inside of bee-hives with mint and honey, or sugar, before the swarm is covered with [the hive], as it is supposed to attach them to the new hive.
> —Phillips, *History of Cultivated Vegetables*, 1: 347

Nasturtium, Indian Cress, Yellow Larkes Heeles. *Tropaeolum majus*

Among hot herbs such as radishes and mustards, wild and cultivated, nasturtiums were particularly appealing in the garden for culinary use and beauty. Their popularity for garnish and flavor has resurged among twenty-first-century cooks. Many agree with the ancient herbalists that nasturtium "is of so great beauty and sweetnesse withall, that my Garden of delight cannot bee unfurnished of it." Today's adventurous diners enjoy not only the leaves but also blossoms and seedpods "because the taste is somewhat sharpe agreeing thereunto."[51]

> When the plants are come up about six inches high, they should have sticks placed for them to run up, for these plants are of the running or climbing kind.
>
> —Mawe, *Every Man His Own Gardener*, 104

> The plants should have pretty long bushy sticks put to them; and four or five of them will bear a great quantity of pods.
>
> —Cobbett, *American Gardener*, 133

> There are two Varieties of this, one with a deep Orange-coloured Flower inclining to red, and the other with a pale yellow Flower . . . and one with double Flowers.
>
> —P. Miller, *Gardeners Dictionary*

> The flowers, being of so excellent a colour for candlelight, are often used to garnish dishes.
>
> —Phillips, *History of Cultivated Vegetables*, 2: 3

> The aromatic buds of nasturtium flowers will also improve the flavour of a mixed salad.
>
> —Von Rumohr, *Essence of Cookery*, 164

> The large Nasturtium is very deserving of cultivation, as well on account of the beauty of its large and numerous orange-coloured flowers, as their excellence in sallads, and their use in garnishing dishes. The green berries, or seeds of this plant, which it produces abundantly, make one of the nicest pickles that can possibly be conceived; in the estimation of many, they are superior to capers.
>
> —M'Mahon, *American Gardener's Calendar*, 318

From the early European herbals to nineteenth-century America we find agreement that nasturtium seedpods do indeed make "one of the nicest pickles." English cookery authors, as E. Smith, typically included at least one receipt. Virginian Mary Randolph documented the method as transplanted to America: "Where capers cannot in garnishing dishes be obtained, pickled nasturtiums [seedpods] make a very good substitute." The simplest of pickling methods comes to us from *The Kentucky Housewife* a century later. Note the evolution of this receipt from Smith in England to Bryan in Kentucky:[52]

> ***To pickle Nasturtium Buds***
>
> Gather your little knobs quickly after your blossoms are off; put them in cold water and salt for three days, shifting them once a day; then make a pickle (but do not boil it at all) of some white-wine, some white-wine vinegar, eschalot [shallot], horse-radish, pepper, salt, cloves and mace whole, and nutmeg quartered; then put in your seeds and stop them close; they are to be eaten as capers.
>
> —Smith, *Compleat Housewife*, 106

FIG. 27. "Nasturtium indicum"

Robert Sayer, *The Florist*, 1760

Naturtians

It is the seeds of the nasturtians that are used for pickling; gather them when full grown, before they get hard, put them in a jar of plain, cold vinegar, close it with a cork, and set it in a cool, dry place; it is not necessary to add salt or spices, as they keep well without.

—Bryan, *Kentucky Housewife*, 182

Nasturtian Sauce

Having drawn your butter in the usual manner [melted and thickened with a little flour], make it as thick as you desire it with pickled nasturtian seeds, add a little of the vinegar in which they were pickled, a small portion of pepper, and serve it warm. It is a fine sauce for boiled mutton.

—Bryan, *Kentucky Housewife*, 167

A most intriguing note regarding the nasturtium is that of English garden historian Henry Phillips (1822), who suggested a lovely excuse to relax in the garden at dusk. According to the daughter of Linnaeus, the Swedish botanist who devised our botanical naming system, nasturtium blossoms "have been observed to emit electric sparks towards evening. . . . It is seen most distinctly with the eye partly closed." Others have claimed to observe this phenomenon from the leaves as well as the blossoms. One wonders if a cordial glass of Hungary water or perhaps claret with a sprig of borage accompanied this crepuscular observation of nasturtiums. At any rate, this is an enchanting pretext for an evening garden party.[53]

Nep.

See **Catmint**.

Onions. *Allium cepa* and other *Allium* species

Phillips' *History of Cultivated Vegetables* proclaimed that the onion "has been the common seasoning for meats with most nations from the earliest time to the present; from the table of Majesty, to that of the peasant." This comes as no surprise, for it still rings true today. Evelyn's *Acetaria* described onions "being eaten crude and alone with Oyl, Vinegar, and Pepper, we own them in Sallet, not so hot as Garlick, nor at all so rank."[54]

An honest laborious Country-man, with good Bread, Salt, and a little Parsley will make a contented Meal with a roasted Onion.

—Evelyn, *Aceteria*, 156

Many people shun onions on account of the strong disagreeable smell they communicate to the breath: this may be remedied by eating walnuts, or a few raw parsley leaves.

—Phillips, *History of Cultivated Vegetables*, 2: 21

One must agree: "To describe all the varieties of the Allium family, we fear, would be over-seasoning the work with onions." Nearly every savory dish incorporates one or more elements from the onion tribe: chives, leeks, shallots, scallions, onions, and so on. Chives were likely to be found in a basic dooryard garden along with others of the smaller onions. Onions, requiring more space to produce an annual supply, were likely grown in the truck garden. Garlic was considered more medicinal than culinary.[55]

Preserving onions is an easy matter. . . . Any dry airy place will, therefore, do. They should not be kept in a warm place. . . . The neatest way is to tie them up in ropes; that is to say, tie them around sticks or straight straw, with matting [or soft string or strips of white oak].

—Cobbett, *American Gardener*, 135, 121

The professional medical community generally considered onions "rather as articles of food than of medicine" and warned of possible side effects—flatulence, thirst, headache, and turbulent dreams. These warnings followed those passed down from the ancient herbalists: "The Onion . . . causeth head-ache, hurteth the eyes, . . . dulleth the senses, ingendreth windinesse, and provoketh overmuch sleepe, especially being eaten raw." Medical benefits were supposed to derive from application rather than internal use.[56]

Clergyman John Wesley was a proponent of simple kitchen remedies. His home medical guide, popular in England and America, proposed an application of raw onion for treating a plethera of complaints ranging from earache, burn, and wasp sting to malarial fever, hemorrhoids, and cancer along with nausea and baldness:

To Stop Vomiting

Apply a large onion, slit across the grain, to the pit of the stomach. [Wesley noted this remedy "Tried."]

—Wesley, *Primitive Remedies*, 126

To Cure Baldness

Rub the part morning and evening, with onions, till it is red, and rub it afterwards with honey.

—Wesley, *Primitive Remedies*, 33

See also **Chives** and **Garlic**.

Parsley

Common Parsley *Petroselinum crispum* var. *neapolitanum*

Curled Parsley *Petroselinum crispum* var. *crispum*

Hamburg, Large-Rooted Parsley *Petroselinum crispum* var. *radicosum*, *P. crispum* var. *tuberosum*

If we could choose only one culinary herb for our *Curious Garden* it would be parsley. Our examination of eighteenth- and early-nineteenth-century sources, European and American, showed parsley to have been the most popular potherb throughout the period. Eighteenth-century French receipts indicated a preference for little more than parsley and chives for their sauces or broths rather than the complicated flavors we expected.

The French . . . speak of us [English, Scots] as a nation "with twenty religions and only one sauce"—parsley and butter, by the way, is this national relish—and unquestionably English cookery, like English manners, has ever been much simpler than that of our neighbours.

—Dods, *Cook and Housewife's Manual*, 227–28

The seed goes nine times to the devil before it comes up, alluding to the length of time it lies in the ground before it germinates, which is generally six weeks.

—*A Treatise on Gardening*, 18–19

There are two kinds of parsley, the common plain, and curl-leaved, but the latter is greatly to be prefered . . . much admired for garnish to dishes.

—Squibb, *Gardener's Calendar*, 9

[Parsley] does indeed give an attractive colour to any meat dish whilst also being compatible with a host of vegetables. It is important not to overcook parsley ... the principle to remember is that parsley will be tasty as long as it retains its beautiful green colour.
—Von Rumohr, *Essence of Cookery*, 162

Parsley ... excellent with bread and butter in the spring.
—Simmons, *American Cookery*, 43

Pickled Parsley is quite a convenient article; it answers for seasoning gravies at a time when fresh parsley cannot be obtained.
—Bryan, *Kentucky Housewife*, 181

The leaves are pleasant in sauces and broth, in which besides that they give a pleasant taste, they be also singular good to take away stoppings, and to provoke urine; which thing the roots likewise do notably performe if they be boiled in broth; they be also delightfull to the taste, and agreeable to the stomacke.
—Gerard, *Herball*, 1014

Parsley also had a place in domestic medical practice. In 1736 Virginia, parsley tea was recommended as the proper beverage for treating kidney stones and suppression of urine, following the advice of earlier herbals. And a century later in North Carolina, parsley root tea was William Lenoir's "Cure for Suppression of Urine."[57]

The Signs of this Complaint [suppression of urine] are too plain to any one who has his Feeling ... you should add Parsley-Tea to your Drink, sweetened with Syrup of Violets ... [or] Drink Parsley Tea, or Cider, with a Plantain Leaf boiled in it.
—Tennent, *Every Man His Own Doctor*, 52–53

Parsley Stone-break.... It cleanses the Kidneys and Bladder of Stone and Gravel.
—*The South Carolina and Georgia Almanack for the Year of Our Lord 1758*, Charleston: John Tobler, 1758

[For] a bruise ... apply a plaster of chopped parsley mixed with butter.
—Wesley, *Primitive Remedies*, 36; also Thompson, Commonplace book, 95

[Poultice for] wounds occasioned by a rusty Nail [Apply] parsley boiled in small Beer with a piece of Butter.
—Unknown author, Herbal and household book

Parsley roots were employed through this time period as well for both culinary and medicinal uses. German broths were customarily seasoned with parsley roots in a "medley" of root vegetables. A nicely seasoned broth of "whatever [meats] you mean to make your bouillon" was discovered among receipts identified as German in Sarah Rutledge's *Carolina Housewife*.

Hamburg Parsley
[Parsley roots] are much of the nature of the Herb [leaves].... They are also good ... for cleansing the Viscera, and particularly the Kidneys.
—Quincy, *English Dispensatory*, 166

These roots are used mainly to season soups and sauces, or they may be combined with medleys of other vegetables. Used alone, their flavour would become rather tedious.
—Von Rumohr, *Essence of Cookery*, 147

Their large root like a parsnep is esteemed by many persons.
—Gardiner and Hepburn, *The American Gardener*, 37

Bouillon

[Boil four pounds of preferred meat and bones for two hours]; put in an onion, a leek, a parsley-root, a carrot, a piece of celery, a leaf of white cabbage, a parsnip, two laurel-leaves [bay], four grains of whole pepper, four cloves, a piece of dried ginger, and two table-spoonsful of salt. These should boil together for one or two hours, until quite tender. . . . It should be strained; the meat left in the sieve, may . . . answer for croquets. German Receipt.

—Rutledge, *Carolina Housewife*, 38–39

[Parsley Roots are] so near the Taste and Flavour of the Fennel Roots, as hardly to be distinguished from them.

—Quincy, *English Dispensatory*, 166

Squibb and M'Mahon recommended parsley for "a neat edging" in their guides for American gardeners—a charming touch worth reviving in our gardens today. And to top it off, Miller's *Gardeners Dictionary* tells us, "Ancients [Greek athletes] made Crowns of it to adorn the Head."[58]

Parsley seed may be sown . . . along the edge of the quarters, or borders of the garden: it will make a useful and also a neat edging . . . especially the curled parsley.

—M'Mahon, *American Gardener's Calendar*, 191

Pennyroyal, Pudding-Grass. *Mentha pulegium*
American Pennyroyal. *Hedeoma pulegioides*

Pennyroyal illustrates the challenge of identification. John Gunn's American *Domestic Medicine* does nothing to solve the puzzle: "There is no description necessary of this plant, as it grows in almost every garden. . . . I will also mention that, the American Pennyroyal is a different plant from that known in England by the same name." So, which pennyroyal did Gunn find growing in frontier gardens? Was it the native *Hedeoma* or the official mint species introduced from Europe? And to which genus were Ellis, Wesley, Byrd, and Kalm referring? Here we have the confusion of common names, which could have been solved by using the taxonomic name. Additional uncertainty arises from Old World versus New World usage and the availability of cultivated versus native plants.[59]

Pennyroyal is toxic and taken internally can lead to serious illness or death. Nevertheless, there is an alarming tradition of pennyroyal in puddings:

The whole kernels of oats, called grotes . . . with milk, butter, spice, and pennyroyal, make oatmeal puddings; but some put to them suet, raisins, &c.

—Ellis, *Country Housewife's Family Companion*, 271

For chin-cough, or hooping cough . . . give a spoonful of juice of pennyroyal, mixed with brown sugar candy, twice a day.

—Wesley, *Primitive Remedies*, 41

Caution: Today we know pennyroyal should be avoided in both food and medicine; the above pudding receipt and cough remedy are strictly for historical interest. By 1802 we find pennyroyal "less frequently used now than formerly," and for good reason. Pennyroyal should never be ingested.[60]

[Ticks and seed ticks, or chiggers] will attack no Part that is rubb'd with the Juice of that fragrant Vegetable.

—Byrd, *Histories*, 274

FIG. 28. Pennyroyal

William P. C. Barton,
Vegetable Materia Medica, 1825

I was likewise told that on feeling a pain in any limb, this plant, if applied to it, would give immediate relief.

—Kalm, *Travels*, 1: 103

If a garland of pennyroyal were worn it would not only ease the head from pain, but that it would preserve the brain from disorders.

—Phillips, *History of Cultivated Vegetables*, 1: 349

Pennyroyal was a favored herb in Tennent's *Every Man His Own Doctor*: a plaister for pleurisy or epilepsy, a tea for fevers or "vapours and hysterick fits."[61]

Peppers: Cayenne, Capsicum, Indian Pepper, Spanish Pepper, Red Pepper.

Capsicum annuum

While most herbs and traditions came to America by way of Europe, capsicum peppers and the enjoyment of their heat and flavors migrated along with enslaved Africans from the Caribbean to southern American seaports, particularly Savannah and Charleston. Multiple common names for peppers are thoroughly confusing, as in the Kalm account below. Although cayenne was sometimes mistakenly called "Guinea pepper," the true Guinea pepper is not the capsicum but the berry of a native West African tree, the wild melegueta. Both types of peppers were similarly enjoyed in traditional foods of Africa and the West Indies.

A complex cultural transfer was evident as South Carolina plantation mistress Eliza Lucas Pinckney sent along "Cayan" to season the sea turtles she attempted to ship alive to friends and relatives in England. English gardeners became acquainted with the cayenne through exposure from the American colonies, or alternatively by way of Spain. *The Gardener's Calendar for South Carolina, Georgia and North-Carolina* covered "Peppers of all Sorts" for cultivating in southern gardens.[62]

Guinea pepper, Capsicum (red) is put to a roasted or boiled piece of meat or mixed with broth. Cucumbers are pickled with it. Pods pounded whilst tender, mixed with salt and preserved in a bottle to be strewed over meat or fish.

—Kalm, *Travels*, 1: 42

Most of the Sorts of Capsicum are Natives of both the Indies; but they have been brought to Europe from America, where they abound in all the Caribbee Islands, and are by the Inhabitants greatly used in all their Sauces, but especially by the Negroes. . . . In Spain and Portugal . . . they are used for the same Purposes, as in America; but in England, they are chiefly cultivated for Ornament, being rarely used for Sauce, or in Medicine; though it is much used in both, in several other Countries.

—P. Miller, *Gardeners Dictionary*

Take ripe chillis, and dry them perfectly in the sun, or before a slow fire, taking care not to scorch them, which would in a great degree spoil their flavor. Grind or pound them to a perfect powder, mixing with them one eighth of their quantity of fine salt. Put it up in small dry bottles, securing the corks with leather. It will keep its strength, and a very small portion of it will give a fine flavor to gravies, soups, &c.

—Bryan, *Kentucky Housewife*, 161

The seed-pods of these plants are much esteemed for pickling.

—Mawe, *Every Man His Own Gardener*, 105

FIG. 29. Pepper, Capsicum

Basilius Besler, *Hortus Eystettensis*, 1613

It is a custom in some places to pickle the green Pods of *Capsicum Indicum* with their Cucumbers, which will contribute to make them much hotter or warmer to the Stomach, and promote digestion in cold Constitutions.
—Bradley, *Country Housewife*, 1: 120–21

Spanish pepper is a capsicum which . . . surpasses all other spices in its warming and inflammatory effects. Cayenne pepper, which has recently become so popular, is prepared from one capsicum variety. . . . A less hot variety of Spanish pepper . . . is used, when still green, to flavour pickles of all kinds. If used with discretion, a few fresh green chilli peppers will improve the flavour of pickled cucumbers without being harmful to the blood. The Italians and the English are in the habit of pickling large jars full of chillies and their habit of using two or three to garnish meat broths can be most agreeable.
—Von Rumohr, *Essence of Cookery*, 166–67

It is said to die [dye] or colour like Saffron; and being received in such sort as Saffron is usually taken, it warmeth the stomacke, and helpeth greatly the digestion of meates.
—Gerard, *Herball*, 366

In *Professor Cullen's Treatise of the Materia Medica* Barton stated that capsicum "is now [1812] very universally employed as a condiment, but has hardly yet had any place in medicinal prescription." Domestic medical advice on the southern frontier in America suggested otherwise:[63]

Pain in the Hip David Hickerson 23 Feby 1815
To 2 spoonfulls of hogs lard, 2 spoonfulls pulverized red pepper & 2 quarts chamber Lye [urine] & boil away till it becomes a Salve & apply it to the part affected.
—Lenoir, Medical memorandum

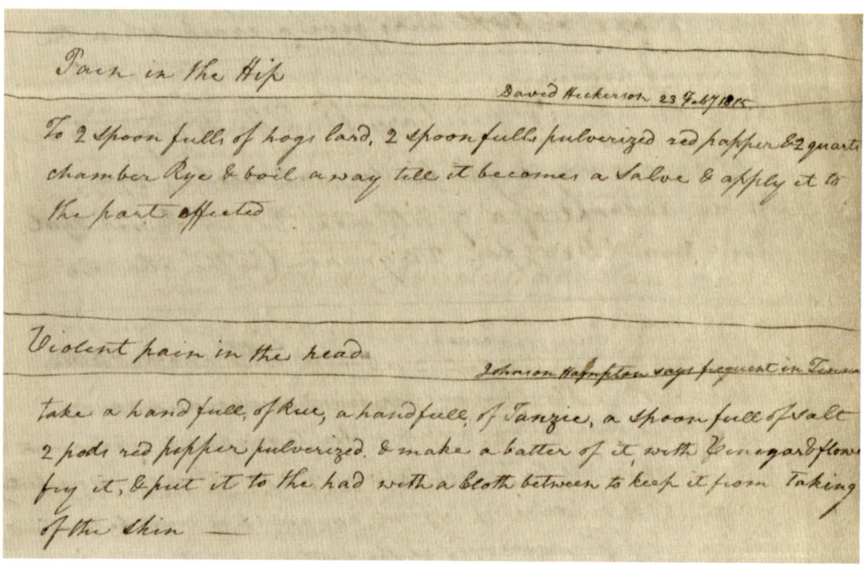
Pain in the Hip
David Hickerson 23 Feby 1815
To 2 Spoonfulls of hogs lard, 2 spoonfulls pulverized red pepper & 2 quarts chamber Lye & boil away till it becomes a Salve & apply it to the part affected

Violent pain in the head
Johnson Hampton says frequent in [illegible]
take a hand full of Rue, a hand full of Tanzie, a Spoon full of Salt 2 pods red pepper pulverized & make a batter of it with Vinegar & flour fry it, & put it to the had with a cloth between to keep it from Taking off the Skin —

FIG. 30. William Lenoir's remedies for "Pain in the Hip" and "Violent Pain in the Head"

William Lenoir, Medical memorandum, North Carolina, 1801–39

Radish. *Raphanus sativus*

Radish leaves, blossoms, and seedpods were enjoyed as well as the roots. Squibb's *Gardener's Calendar for South Carolina, Georgia and North-Carolina* and Philip Miller's *Gardeners Dictionary*, among other gardening guides, recommended planting radishes in every month of the year to ensure a continuous supply. Reviving the practice of interplanting radishes of several sorts with other herbs saves garden space and adds visual interest in today's garden as it did two centuries ago.

> Many of the Gardeners near London sow Carrot Seed with their early Radishes, so that, when their Radishes are killed, which often happens, the Carrots will remain, for the Seeds of Carrots commonly lie in the Ground five or six Weeks before they come up . . . but, when both Crops succeed, the Radishes must be drawn off very young. . . . It is also a constant Practice with these Gardeners to mix Spinach Seed with their latter Crops of Radishes, so that when the Radishes are drawn off, and the Ground cleaned between the Spinach, it will grow prodigiously.
>
> —P. Miller, *Gardeners Dictionary*

Note how the following advice for late-eighteenth-century and early-nineteenth-century American gardeners echoed Miller.

> [In January or February:] There may be mixed and sown with the radishes a sprinkling of carrot seed; for if the radishes should fail, the carrots may succeed; and if both should succeed, there will be a double advantage . . . or, instead of carrots, you may sow a little lettuce seed or round spinage, and when the radishes are gone off these will come in.
>
> —Squibb, *Gardener's Calendar*, 12; M'Mahon, *American Gardener's Calendar*, 123

Radishes may be preserved in sand, as Carrots are; till the spring.

—*Treatise on Gardening*, 48

Radish is sauce for cloyed stomacks, as Capers, Olives, and Cucumbers: cast the seeds all summer long here and there, and you shall have them alwayes young and fresh.

—W. Lawson, *Country House-wives Garden*, 82

[Radish roots] are eaten alone with Salt only, as carrying their Peper [*sic*] in them.

—Evelyn, *Aceteria*, 159

The black and white Spanish Radishes are commonly cultivated for medicinal Use, though there are some Persons who are very fond of them for the Table.

—P. Miller, *Gardeners Dictionary*

The [salad] dishes are garnished with slices of black spanish radish.

—Evelyn, *Directions for the Gardiner*, 119

[French peasants] make great use of [radishes] in their aliments; they roast them under the ashes, and put them also into soup, to which they give an agreeable taste. Cattle are also fed with them during the winter months.

—Phillips, *History of Cultivated Vegetables*, 2: 208

We are informed, that the leaves were carefully taken off, in order to increase the size of the roots.

—Phillips, *History of Cultivated Vegetables*, 2: 108

To Prevent the Bite of a Viper

Rub the hands with the juice of radishes.

—Wesley, *Primitive Remedies*, 124

[To remove] Warts

Rub them daily with a radish.

—Wesley, *Primitive Remedies*, 128

Radish seedpods were pickled akin to nasturtium pods. Compare the following mid-eighteenth-century English receipt with the South Carolina method from a century later. It is interesting that among mid-eighteenth-century English authors Smith included garlic rather than the more typical horseradish and mustard seed combination.

> *To pickle Pods of Radish*
> Gather the youngest Pods, and put them in water and salt twenty-four hours; then make a pickle for them of vinegar, cloves, mace, whole pepper; boil this, and drain the Pods from the salt and water, and pour the liquor on them boiling hot: Put to them a clove of garlick [or more commonly, horseradish and mustard seed] a little bruised.
> —Smith, *Compleat Housewife*, 110

> *To pickle Radish Pods [a century later in South Carolina]*
> Salt and dry them in the sun; when quite dry, throw them into vinegar which has been boiled, but is quite cold.
> —Rutledge, *Carolina Housewife*, 182

Rose.
Apothecary, Red. *Rosa gallica* 'Officinalis'
Blush. *Rosa chinensis*
Damask. *Rosa × damascena*
Musk. *Rosa moschata*
Provence. *Rosa × centifolia*
Eglantine, Sweetbriar, Wild. *Rosa rubiginosa*

Our ancestors delighted in roses for their beautiful blossoms as well as their delightful scents. The enchanting flavor of rose in baked goods and puddings is worthy of rediscovery.

> The Rose doth deserve the chiefest and most principall place among all floures whatsoever, beeing not onely esteemed for his beautie, vertues, and his fragrant and odoriferous smell.
> —Gerard, *Herball*, 1259

Flower waters were the flavorings of choice before vanilla captured our taste buds. Instead, eighteenth-century receipts called for rose or orange flower water (and later peach). Flower waters are a product of distillation, and most kitchens were not equipped with the necessary small still. *The Virginia Housewife* and *The Kentucky Housewife* offered a lovely alternative: a simply made infusion of rose petals in brandy requiring no more equipment than a ceramic jar. An annual supply of rose brandy is easily prepared when scented roses bloom. This tincture is a much nicer and gentler flavoring agent than today's commercial rose waters. Once tried, you are likely to adopt rose brandy as your new favorite flavoring for pound cake or rice pudding or even pumpkin pie.

Old scented rose varieties are available. Many can be grown successfully without pesticides, ready for restocking your own annual rose brandy supply. Be aware that in early sources, as in *The Virginia Housewife*'s "Rose Brandy" (below), "rose leaves" most likely indicates flower petals rather than foliage. On the other hand, the actual green leaves of the eglantine rose or sweet brier are also fragrant. *The Kentucky Housewife*'s "Rose Brandy" is similar to Mary Randolph's with an interesting embellishment: "To make the liquid more odoriferous, you may add some fragrant clove pink leaves, sweet williams, &c."[64]

FIG. 31. Roses

Three varieties particularly noted for medicinal qualities: officinal rose or red rose; dog-rose or hip tree; hundred-leaved rose or damask rose.

Nicholas Culpeper, *Complete Herbal*, 1826

Rose Brandy

Gather leaves [petals] from fragrant roses without bruising, fill a pitcher with them, and cover them with French brandy; next day, pour off the brandy, take out the leaves, and fill the pitcher with fresh ones, and return the brandy; do this till it is strongly impregnated, then bottle it; keep the pitcher closely covered during the process. It is better than distilled rose water for cakes, &c.

—M. Randolph, *Virginia Housewife*, 214

The rose fruits, or hips, especially those of the eglantine or wild rose, were enjoyed in tarts and pastes. Any rose hips may be used, although flavors differ. Larger ones are easier to handle.

> Children with great delight eat the berries thereof [eglantine, sweet brier] when they be ripe, make chaines and other prettie gewgawes of the fruit: cookes and gentlewomen make Tarts and such like dishes for pleasure thereof.
>
> —Gerard, *Herball*, 1269

> The [wild rose] fruit when it is ripe maketh most pleasant meats and banqueting dishes, as tarts and such like; the making whereof I commit to the cunning cooke, and teeth to eat them in the rich mans mouth.
>
> —Gerard, *Herball*, 1271

To Make a Tart of Hipps

Slyce them & cut out the kernells, & seeth them in white wine, & when they be soft, streyne them as thick as you can; Then season them with sugar, cinnamon, ginger, rose water, then put this in crust made of a tart fashion.

—*Martha Washington's Booke*, 97

William Lenoir's personal medical memoranda from early-nineteenth-century western North Carolina suggested wild rose roots as efficacious for diarrhea, kidney problems, and old sores. These uses echoed the general virtues of roses outlined two centuries earlier in Gerard's *Herbal.*

> The leaves [petals] of the [musk rose] floures eaten in the morning, in manner of a sallad, with oile, vineger and pepper, or any other way . . . purge very notably the belly of waterish and cholericke humors.
>
> —Gerard, *Herball*, 1268

> [For hoarseness] . . . a tea-spoonful of conserve of roses every night.
>
> —Wesley, *Primitive Remedies*, 80

> In the [apothecary] shops are prepared a conserve, an infusion, a honey, and a syrup of this flower.
>
> —*Edinburgh Dispensatory*, 266

> The practice has been to prepare [rose] syrup with honey rather than sugar; but we cannot find any advantage to arise from the honey.
>
> —W. Barton, *Cullen's Treatise*, 2: 26

> Take of red roses, half a pound, strong vinegar, half a gallon. Infuse in a close vessel for several weeks, in a gentle heat; then strain off the liquor. This is chiefly used as an embrocation [liniment] for headaches, &c.
>
> —Thomson, *Family Physician*, 482

See **A Salmagundi of Methods** for conserve of red roses, syrups, honeys, and other flower preparations.

Rose Campion, Catchfly, Gardener's Eye.

Silene coronaria (syn. *Lychnis coronaria*)

The 1806 *American Gardener's Calendar* documented continuing interest in cultivating the lovely rose campion. When introducing this showy flower into your garden, be aware that rose campion can be invasive.

> Because the leaves thereof be soft, and fit to make weekes [wicks] for candles . . . lightgiving floures: and therfore they were called the Gardners [*sic*] Delight, or the Gardeners Eye.
>
> —Gerard, *Herball*, 468

> It hath the name of Catch . . . because the stalkes in the hot Summer dayes have a certaine viscous or clammy humour upon them, whereby it easily holdeth . . . whatsoever small thing, as Flies, &c. lighteth upon it.
>
> —Parkinson, *Paradisi*, 256

> The seed of Wilde Campion beaten to pouder and drunke . . . is good for them that are stung or bitten of any venomous beast.
>
> —Gerard, *Herball*, 471–72

Rosemary. *Rosmarinus officinalis*

Explorer John Lawson found rosemary already growing in the Carolinas at the beginning of the eighteenth century. In 1760 a wagonload of goods from the Cape Fear region arrived in the North Carolina Moravian settlement of Bethabara with rosebushes and rosemary. Rosemary seems to have been less favored for culinary use in the eighteenth century than in the preceding century or today, but it was grown and valued for a variety of other reasons. In 1660 rosemary's use was "much in meats, more in Physick [medicine], most for Bees"; in mid-1760 North Carolina, "Hungary Water is made from it"; in receipts from early-nineteenth-century Germany we read of rosemary "used in stuffing poultry."[65]

> *[In "December's Husbandry"]*
>
> Go look to thy bees, if the hive be too light,
> Set water and honey, with rosemary dight;
> Which set in a dish full of sticks in the hive,
> From danger of famine ye save them alive.
>
> —Tusser, *Good Husbandry*, 64

> Rosemary gives out its virtues completely to rectified spirit, but only partially to water. . . . It is a principal ingredient in what is known by the name of Hungary-water.
>
> —Thomson, *Family Physician*, 416

> ***To make Hungary Water***
>
> Take four ounces of rosemary-flowers, and a pint of spirits of wine [brandy], infuse it twelve hours, and draw it off in a glass still.
>
> —Smith, *Compleat Housewife*, 272

A most appealing attribute is often quoted from Shakespeare's *Hamlet*: "There's rosemary, that's for remembrance." Early herbals detailed associated traditions of rosemary at "weddings, funerals, &c. to bestow among friends." These sweet and affecting traditions continued through several centuries and deserve to be revived. The following extracts may send us out to the garden to weave a crown of rosemary, or more certainly to toss a sprig of rosemary into a glass of wine.[66]

[Rosemary for] comforting the brain and strengthening the memory . . . emblematical of fidelity in love: it was therefore woven into coronets, and worn at weddings; and perhaps on the same principle at funerals . . . in some parts of England it is still [in 1822] distributed among the company, who generally throw the sprigs into the grave.

—Phillips, *History of Cultivated Vegetables*, 2: 136–37

[Rosemary] Flowers, a little bitter, are always welcome in Vinegar; but above all, a fresh Sprig or two in a Glass of Wine.

—Evelyn, *Aceteria*, 160

[When rendering lard from pork fat,] a good housewife commonly lets a sprig (two or three) of rosemary be amongst the fat in melting, for giving the lard an agreeable flavour.

—Ellis, *Country Housewife's Family Companion*, 131; echoed by Rundell, *New System of Domestic Cookery*, 66

The [rosemary] flowers and conserves made of them are singularly good . . . to expel the contagion of the pestilence; and . . . to burn the branches of it in infectious times. . . . The dried herb burnt is good to sweeten the air, and correct noxious and disagreeable smells.

—Phillips, *History of Cultivated Vegetables*, 2: 136

[As therapy following a palsey (stroke)] beat Rosemary in a mortar, and make a little Ball of it, which you must roll and work about in your Hands continually, renewing the same every Day.

—Tennent, *Every Man His Own Doctor*, 27

For the Cramp

Take of rosemary-leaves, and chop them very small, and sew them in fine linen, and make them into garters, and wear them night and day.

—Smith, *Compleat Housewife*, 323

To make hair grow. Wash it every night with a strong decoction of rosemary. Dry it with flannel.

—Wesley, *Primitive Remedies*, 76

Rue, Herbe Grace. *Ruta graveolens*

The *Records of the Moravians in North Carolina* noted that rue was grown in local gardens and specifically in the medical garden. Gunn's *Domestic Medicine* described rue as "valuable to weak and hysterical constitutions." Rue was touted in Old World sources as a remedy for a multitude of ills and to prevent infections when worn "as a nosegay in our visits to those who suffer under contagious diseases."[67]

Caution: Be aware that rue may irritate the skin or trigger blistering, despite the old herbalist's perception of "many good properties whereunto Rue serveth." Rue remained "extreemly common" in English gardens at the dawn of the nineteenth century.[68]

The many good properties whereunto Rue serveth . . . in former times caused the English name of Herbe Grace to be given unto it.

—Parkinson, *Paradisi*, 530

[Rue is] too strong for mine Housewifes pot, unless she will brew Ale therewith, against the Plague.

—W. Lawson, *Country House-wives Garden*, 82

The flavour of rue is so strong that it is not generally popular as an ingredient in herbal spinach, herb soups or in herb sauces, either hot or cold.

—Von Rumohr, *Essence of Cookery*, 163

We must wonder about William Lenoir's remedy for "violent pain in the head." At least he included a warning that this application of rue, tansy, and red pepper may take off the skin:

> *Violent Pain in the Head*
> Take a handfull of Rue, a handfull of Tanzie, a spoonfull of salt 2 pods red pepper pulverized & make a batter of it with Vinegar & flower [flour] fry it, & put it to the had [head] with a Cloth between to keep it from taking off the Skin.
> —Lenoir, Medical memorandum (see figure 30)

> [Rue] killeth and driveth forth the Worms of the Belly, if it be drank after it is boiled in Wine to the half, with a little Honey.
> —Culpeper, *English Physician*, 291

Saffron, English Saffron, Fall Crocus.

Crocus sativus

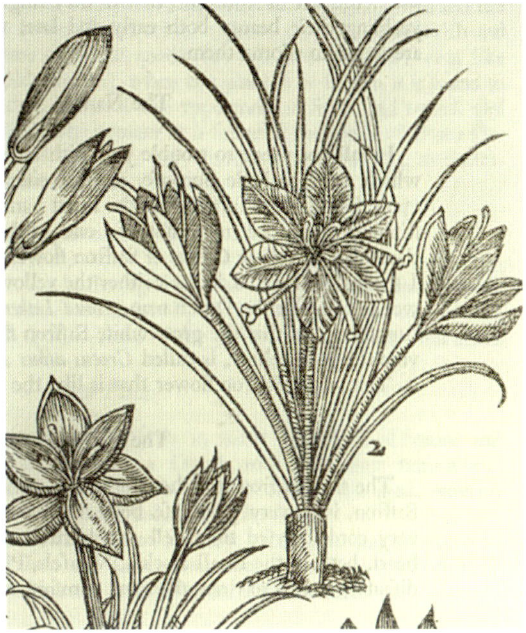

FIG. 32. "The true Saffron"

John Parkinson, *Paradisi in Sole*, 1613

The early herbals tell us saffron was "used in meates and medicines" and as "a notable Remedy for the Yellow-Jaundice"; but "hurtful to the braine . . . by too much using of it. . . . But the moderate use of it is good for the head, and maketh the senses more quicke and lively." Saffron's yellow essence suggested its use against jaundice following the Doctrine of Signatures. This practice continued into the southern colonies, with Eliza Lucas Pinckney listing "An Excellent Receipt for the Jaundice" utilizing pennyroyal, yarrow, and saffron.[69]

We were surprised to find saffron mentioned in garden lists from the eighteenth-century Carolinas and the middle colonies: "Many of the inhabitants plant saffron; but it is not so good or so strong as the English and French variety." Only the red stigma from inside the autumn crocus bloom is harvested for the desired scent, flavor, and color. "When the Saffron comes to flower, the Blossoms must be gathered very early in the Morning. Because . . . the Stile or Pistilum, which is the pure Saffron, shrinks at the approach of the Sun. . . . As they gather the Saffron, they put it between Sheets of White Paper . . . and dry it." The German culinary writer Von Rumohr thought saffron overrated.[70]

> Saffron is an absurd seasoning which is esteemed in some countries just for the sake of its beautiful yellow colour. Its flavour is quite weak, though not unpleasant.
> —Von Rumohr, *Essence of Cookery*, 162

[To prevent an attack of ague] take twenty grains of powered saffron before the fit, in a glass of white wine.
—Wesley, *Primitive Remedies*, 27

Biles [Boils]
Apply . . . a little saffron in a white bread poultice.
—Wesley, *Primitive Remedies*, 35

[For a yellow food coloring] the bloom of garden saffron will be found very convenient, and the color not inferior to any other yellow. . . . To prepare the saffron, put the blossom, after bruising it a little, in an earthen vessel, cover it well with cold water, cover it, and let it stand a few hours to infuse, until it is a bright yellow; then filter it through a piece of muslin.
—Bryan, *Kentucky Housewife*, 303–4

The watery and vinous tinctures are apt to grow sour, and then loose [lose] their colour also; that made in pure spirits keeps in perfection for many years. Its officinal preparations are, a spirituous tincture and syrup.
—*Edinburgh Dispensatory*, 178

The curd for . . . cheese may be coloured with a little saffron.
—Dods, *Cook and Housewife's Manual*, 507

The women in Ireland formerly dyed their body-linen with saffron, to procure . . . strength of body, and cheerfulness of mind. . . . The lower classes tinge their shirts with saffron to keep off vermin . . . [and] for preventing putrefaction. . . . Saffron affords a yellow colour to the dyers, and a useful one to the painters in their watercolours.
—Phillips, *History of Cultivated Vegetables*, 2: 196–97, 201

Sage. *Salvia officinalis*

"It is called Sage in English from the French word *sage* (wise), from the property it is said to possess of strengthening the memory, and thus making people sage or wise." Long-standing tradition applauded the use of sage as a seasoning and also as a preservative for meats or for prolonging human life. Aside from the fact that sage is tasty in cheese or in butter when added with a light hand, it is amusing to suppose we are fending off the signs of aging while we enjoy our bread with sage butter or cheese.[71]

Sage is much used of many in the month of May fasting, with butter and Parsley, and is held of most much to conduce to the health of mans body.
—Parkinson, *Paradisi*, 478

This valuable garden herb was once supposed by the ancients, to prolong the lives of those who would frequently use it.
—Gunn, *Domestic Medicine*, 385

Sage is of excellent use to help the Memory . . . the Conserve made of the Flowers is used to the same Purpose.
—Culpeper, *English Physician*, 296

It quickens the Senses and Memory, strengthens the Head, Nerves and Joints. . . . Sage used as Tea is very wholesom.
—*The South Carolina and Georgia Almanack for the Year of Our Lord 1758*, Charleston: John Tobler, 1758

To Make a Plain Sage-Cheese
Gather the young Tops of red Sage, and bruise them in a Mortar till you can press the Juice from them; then take

Leaves of Spinach or Spinage, and bruise them likewise, and press out the Juice to mix with the Sage Juice; for the Sage Juice of it self is not of a pleasant green Colour. . . . When this Juice is prepared, put your Rennet to the Milk, and, at the same time, mix as much of your Sage and Spinach Juice with it, as will give the Milk the green Colour you desire. [Then proceed with cheese production.]
—Bradley, *Country Housewife*, 1: 79–80

Sage is very useful both as a medicine, for the headache—when made into tea—and for all kinds of stuffing, when dried and rubbed into powder.
—Child, *American Frugal Housewife*, 36

In 1802 Alexander Thomson published in America an antiquated notion that "sage is endowed with the power of resisting the putrefaction of animal substances." Then, as now, sage commonly flavored sausages. Cooks also favored sage in stuffings for poultry. In the first cookbook written by an American we read: "Sage is used in Cheese and Pork, but not generally approved." In 1830 Tennessee, sage was "too well known, and too much used to require a description."[72]

There are 15 different sorts, but the common or green, and the red are principally cultivated.
—*A Treatise on Gardening*, 52

Sage being clipped spring & Autumn, will make it continue long, & faire.
—Evelyn, *Directions for the Gardiner*, 107

If little branches of common sage or Roman sage are placed between small birds on a spit, they will flavour the meat in a most appropriate manner. Sage can be placed between pieces of fresh eel on a spit, to be thoroughly roasted with the fish. It can also be crisped in clarified butter and used as an accompaniment to a variety of other baked goods.
—Von Rumohr, *Essence of Cookery*, 163

See the herb fritter receipt under **Clary** for an interesting alternative to frying sage, parsley, or comfrey leaves.

Fine herbs such as parsley and sage can be deep fried until really crisp and will make a decorative and wholesome accompaniment to other fried foods. The knack of deep frying these herbs is so elusive that the task is often given to trainee cooks to test their ability.
—Von Rumohr, *Essence of Cookery*, 111

Sage Drink [for illness]
Take a little Sage, a little Baum [balm], put it into a Pan, slice a Lemon, Peel and all, a few Nobs of Sugar, one Glass of White Wine, pour on these two or three Quarts of boiling Water, cover it, and drink when dry.
—Glasse, *Art of Cookery* (1747), 121

The Side-shoots and Tops . . . may be gather'd in the Summer, and dry'd, if design'd for Tea, otherwise they are best taken green from the Plants for most other Uses.
—P. Miller, *Gardeners Dictionary*

The broad leaved sage is preferable for tea.
—*A Treatise on Gardening*, 52

For Sunburn, (smarting). Wash the face with sage tea.
—Wesley, *Primitive Remedies*, 116

In ancient times sage was celebrated as a remedy of great efficacy; but at present [1802] it is considered as of little importance in the Materia Medica.
—Thomson, *Family Physician*, 217

See also **Clary, Clary Sage**.

Savory.
Summer. *Satureja hortensis*
Winter. *Satureja montana*

Savory has long been considered "good for my Housewifes pot and pye." It was customarily included in the faggot (bundle) of sweet herbs when seasoning an eighteenth-century dish. John Lawson found both summer and winter savory already being grown in North Carolina in 1709.[73]

> There be two kindes of Savorie, the one that indureth Winter, and is of long continuance, the other an annual or yearely plant.
>
> —Gerard, *Herball*, 575

> Both these are so well known (being entertained as constant Inhabitants in our Gardens) that they need no Description. . . . [K]eep it dry by you all the Year, if you love yourself and your Ease . . . make Conserves and Syrups of it.
>
> —Culpeper, *English Physician*, 302

> *To make a Pidgeon pye in puff paste*
>
> Take the pidgeons & quarter them & Lay them in a deep dish with good paste under them. . . . Season them with Pepper & Salt, Cloves & mace & quantity of Capers . . . take a good many of the tops of winter Savoury, the yolks of two or three hard Eggs, & a good quantity of butter, put in the Pickle of Capers a Spoonful . . . or Else a Spoonful or two of wine then cover it with very fine paste . . . [and bake] a little above an hour.
>
> —"Jane Randolph her Cookery Book" (1743), in Harbury, *Virginia's Cooking Dynasty*, 322

> Summer-savory is excellent to season soup, broth, and sausages. As a medicine, it relieves the cholic.
>
> —Child, *American Frugal Housewife*, 36

> [Savory] doth marvellously prevaile against winde: therefore it is with good successe boiled and eaten with beanes, peason, and other windie pulses, yea if it be applied to the belly in a fomentation, it forewith helpeth the affects . . . proceeding from winde.
>
> —Gerard, *Herball*, 577

> Rectified spirit [brandy] extracts the whole of [summer savory's] taste and smell.
>
> —*Edinburgh Dispensatory*, 278

> Among the fragrant herbs and shrubs that should grow near the beehive.
>
> —Phillips, *History of Cultivated Vegetables*, 2: 210

Sorrel. *Rumex acetosa*
French Sorrel. *Rumex scutatus*
Sheep Sorrel. *Rumex acetosella*
Woodsorrel. *Oxalis acetosella*

Early Americans enjoyed the pleasant tartness of both sorrel (*Rumex*) and woodsorrel (*Oxalis*), as did the old herbalists. "There be divers kindes of Sorrell . . . some of the garden, others wilde; some great, and some lesser." This acid tang from the garden offered a fine alternative to vinegar or lemon. *The Kentucky Housewife* confirmed the fondness for this herb in America: "Sorrel should be dressed like spinach, and many people think when they are mixed together in equal proportions, the taste of both is much ameliorated."[74]

FIG. 33. Sorrel

Elizabeth Blackwell, *A Curious Herbal*, 1739

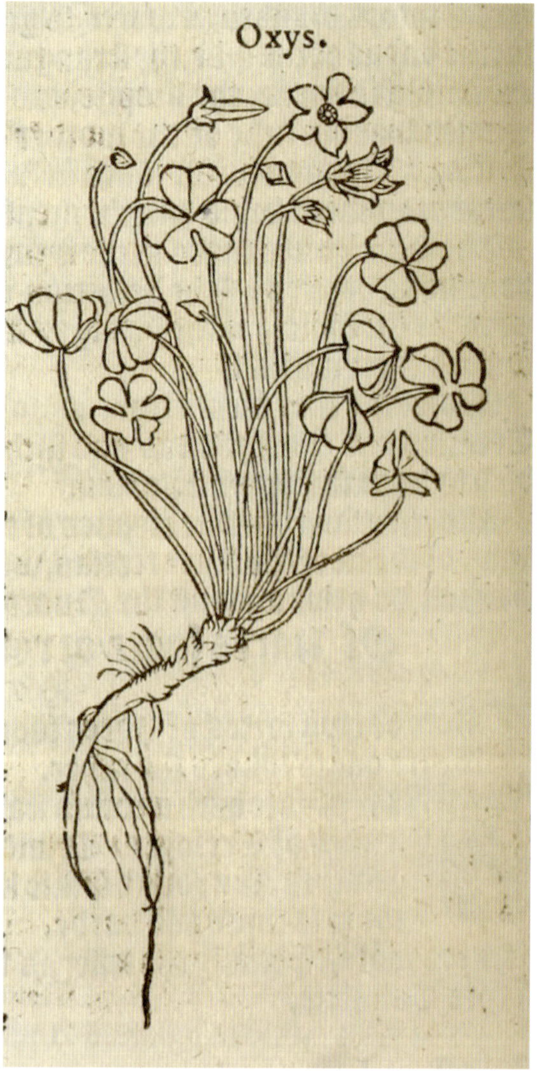

FIG. 34. Oxalis, Woodsorrel

Rembert Dodoens, *A Niewe Herball*, 1578

Our ordinary Sorrel, which grows in Gardens, and also wild in the Fields, is so well known, that it needeth no Description.

—Culpeper, *English Physician*, 314

This is no other than the wild sorrel cultivated.

—Cobbett, *American Gardener*, 143

[Woodsorrel] has passed with many for a Cordial from its Signature, its Leaf resembling a Heart; but such Conceits are not now regarded.

—Quincy, *English Dispensatory*, 216

[Woodsorrel, oxalis] of all Sorrel sauces is the best, not onely in vertue, but also in the pleasantnesse of his taste.

—Gerard, *Herball*, 1202

Sorrel leaves can enhance the flavour of stocks, they can flavour a variety of sauces and can be consumed as a vegetable, either alone, or mixed with other herbs.

—Von Rumohr, *Essence of Cookery*, 160

Pease-soup

Boil the green pease without their shells, when tender strain and hull them through a cullender, then boil the drain'd liquor with parsley, marygolds [petals], lettice, sorrel, and mint, adding butter, some flower [flour], a little pepper, salt, and bits of toasted bread.

—Ellis, *Country Housewife's Family Companion*, 292–93

To Stew Sorrel for Roasts of Veal, Lamb, Fricaneaux [Fricassee], &c.

Wash and simmer it . . . very slowly, and beat it up with a bit of butter, or a little salad-oil. Add cayenne, pepper, and salt. A mixture of spinage and sorrel is dressed as spinage, where sorrel alone might be thought too strong in acid.

—Dods, *Cook and Housewife's Manual*, 215

The housekeepers often cook sorrel broth.

—Fries, *Records of the Moravians*, 2: 570

Oats with Sourleaf to cook

One checks over the oats, then wash them well and soak them in hot water. Take beef broth and put it in a soup pot with onions, whole nutmeg and let this cook a little. Your oats should be added to the broth. Don't add much of it, then add the sourleaf to the broth . . . then added to the oats and a bit of fresh butter, salt it a little and stir in a beaten egg yolk and a little sweet cream and pour it over bread slices.

—Winkler Family recipe

[Sorrel Sauce]

A sauce can be made by boiling freshly chopped sorrel in meat stock . . . and then thickening it with a few egg yolks.

—Von Rumohr, *Essence of Cookery*, 160

[Woodsorrel] beaten with sugar, a very elegant conserve.

—*Edinburgh Dispensatory*, 224

A Posset with Wood-Sorrel

Take new Milk a pound; good fresh Small-beer a pound; White Wine half a pound; and boil them together with 2 handfuls of green Wood-Sorrel a little bruised. This may also be made only by boiling Wood-Sorrel and Milk together, and it makes a pleasant acid Whey when the Curd is taken off; it is a very good Drink in Fevers.

—Quincy, *English Dispensatory*, 696

[Woodsorrel] has a grateful acid taste, approaching nearly to that of the juice of lemons. . . . The principal use, however, of the acetosella [sheep sorrel], is to allay inordinate heat, and to quench thirst; for which purpose, a pleasant whey may be made by boiling the plant in milk.

—Thomson, *Family Physician*, 425

Sorrel was employed as a mordant to set dyes. Curiously, it was also suggested as a stain remover, "commonly used for taking ink-stains out of linen." Stains such as pot black and red clay presented an everyday challenge to housewives.[75]

The French use the juice of sorrel to prepare their hemp and flax for a red dye. . . . [The Dutch and Scots also used sorrel as mordant] to give an excellent black colour to their woollen stuffs, without the use of copperas.

—Phillips, *History of Cultivated Vegetables*, 2: 221, 223

To take out iron-moulds [stains]

Hold the spot stretched wet over a jug of hot water, and rub it with . . . juice of sorrel, and salt; then wash it immediately.

—Dods, *Cook and Housewife's Manual*, 513

Southernwood. *Artemisia abrotanum*

A trusted medicinal herb from the Old World, southernwood was "well known to be an ordinary inhabitant in [English] Gardens." We learn from lists compiled by Moravian surveyor Christian Reuter that southernwood was being grown in North Carolina gardens by 1761. Like other artemisias, southernwood was administered to rid the body of worms. More generally, Gerard's herbal suggested, "Southernwood drunke in wine is good against all venome and poyson . . . strewed about the bed . . . it driveth away serpents."[76]

As Southernwood has been distinguished, however, from Wormwood by all the antient Botanists, and the English Name of Southernwood being well known, I shall continue this Distinction.

—P. Miller, *Gardeners Dictionary*

[It was] given to kill Worms in Children. The Herb bruised and laid to, helpeth to draw forth Splinters and Thorns out of the Flesh . . . likewise killed Lice in the Head. . . . Ashes mingled with Sallad [olive] Oil, helpeth those that have hair fallen, and are bald, causing the Hair to grow again either on the Head or Beard.

—Culpeper, *English Physician*, 318

Leaves when bruised, emit an agreeable Odour. This Sort was formerly used in Medicine, but of late [by 1735] has been expelled [from] the [apothecary] Shops.

—P. Miller, *Gardeners Dictionary*

It is now [1802] rarely used, unless in the way of [a] fomentation.

—Thomson, *Family Physician*, 426

Spinach, Spinage. *Spinacia oleracea*

John Lawson, traveling through the Carolinas early in the eighteenth century, grouped spinach with lettuces, as we might do today. In contrast, early authors considered spinach best when cooked, to become "a most excellent Condiment." Many recommended adding acid—citrus, vinegar, or sorrel—to brighten the flavor. Spinach was also approved in springtime puddings for both flavor and color.

Spinach, of old not us'd in Sallets, and the oftner kept out the better. . . . But being boil'd to a Pult [pulp], and without other Water than its own moisture, is a most excellent Condiment with Butter, Vinegar, or Limon, for all sorts of boil'd Flesh, and may accompany a Sick Man's Diet.

—Evelyn, *Acetaria*, 162

A Fricando [Fricassee] of Beef

Cut a few slices of beef six inches long, two or three wide, and one thick, lard them with bacon, dredge them well [with flour], and make them a nice brown before a brisk fire; stir them half an hour in a well seasoned gravy, put some stewed sorrel or spinage in the dish, lay on the beef, and pour over a sufficient quantity of gravy.

—M. Randolph, *Virginia Housewife*, 41

The seventeenth-century herbalist John Gerard declared: "This herbe, of all other pot-herbs and sallade herbes, maketh the greatest diversities of meats and sallades." *The American Gardener* agreed: "Every one knows how good and useful a plant this is." This opinion has persisted. Spinach—boiled, chopped, seasoned with butter, salt, and pepper—was (and is) also commonly paired with eggs.[77]

Having poached some eggs in the nicest manner, lay them over the top of the spinach, sprinkle over them a spoonful of hot vinegar, with a little black pepper, and send them up warm.

—Bryan, *Kentucky Housewife*, 205

Spinach juice was employed as a green food coloring, especially in boiled or baked puddings such as the traditional spring "tansy" or the basic bread pudding below. More than a century and a half later we find *The Carolina Housewife* and *The Kentucky Housewife* echoing spinach as a food-coloring agent. An early-nineteenth-century German source suggested the green mashed potatoes included here, an attractive and tasty dish in any century.

Cheese is tinged green with juice of spinage; flavoured or slightly tinged with the expressed juice of red sage.

—Dods, *Cook and Housewife's Manual*, 503

To make a green Pudding to Butter

Take a Quart of Cream and boil it, then put in twelve Eggs, yolks and whites well beaten, and one Manchet [small loaf or roll of white bread] grated small, a little salt, beaten Spice and some Sugar [one-half cup sugar is about right]. . . . Then colour it well with some Juice of Spinage, or if you will have it yellow, colour it with Saffron, so boil it in a wet Cloth flowred . . . and serve it in with Wine, Sugar and Butter.

—Wolley, *The Queen-like Closet*, 92

[Potato Puree]

Potato puree can be sieved together with a mixture of fine herbs. One half of the quantity used should be spinach, and the other half sorrel, chervil, parsley, purslane and tarragon. These herbs should be heated in a little concentrated stock so that they they retain their fine green colour. Then chop them, taking care not to loose [lose] their juice. [Add the herbs and liquid to twice their quantity of potato along with a little butter and salt.]

—Von Rumohr, *Essence of Cookery*, 139

Stonecrop, Sengreen, Sedum.

See **Houseleek**.

Tansy. *Tanacetum vulgare*

Caution: Twenty-first-century insight tells us that tansy is toxic and should not be ingested in food or medicine. Be aware that some people experience contact dermatitis from handling tansy.

Tansy's bitterness recommended it to ancient herbalists. Miller's widely consulted *Gardeners Dictionary* documented its popularity: "The common Tansy is greatly used in the Kitchen early in the Spring." Explorer John Lawson and Moravian settlers documented tansy in the Carolinas. In nineteenth-century Tennessee, medical author John Gunn noted that tansy continued to be "most frequently cultivated in gardens, both for culinary and medicinal purposes," among the bitter herbs of springtime. Similarly, Cobbett's *American Gardener* ranked tansy "a perennial culinary and medicinal herb." However, some writers cautioned cooks to limit use of this herb:[78]

> Tansies must be eaten only on Palm and Easter Sundays and their Neighbor Days.
>
> —Culpeper, *English Physician*, 327

> Sparingly mixt with our cold Sallet, and much fitter (tho' in very small quantity) for the Pan, being qualified with the Juices of other fresh Herbs . . . at the entrance of the Spring.
>
> —Evelyn, *Aceteria*, 163

"Tansy" was a common name for an herbal pudding as well, and we find early tansy pudding receipts both with and without tansy as an ingredient. The evolution of receipts suggests the herb tansy had generally fallen from popular use by the end of the eighteenth century. Among tansy puddings are assorted custard or bread puddings and pancakes or omelets colored with springtime leafy greens. A tansy might have been

FIG. 35. Tansie

Elizabeth Blackwell, *A Curious Herbal*, 1739

boiled, baked, or fried like the following tansy from 1685 England.

Healthful springtime greens are worth celebrating in any century. For safety we now know to totally omit tansy and feverfew from the mix. Spinach, sorrel, and violet leaves may be substituted in place of the questionable herbs.

> *[Tansy] Otherways*
> Take a little tansie, featherfew, parsley, and violets stamp and strain them with eight or ten eggs and salt, fry them in sweet butter, and serve them on a plate and dish with some sugar.
> —May, *Accomplisht Cook*, 145

Hannah Glasse's eighteenth-century English cookbook editions contained four tansy pudding receipts; all but one actually called for "sprigs" of tansy. By 1805, however, in Glasse's first American edition we find only one receipt remaining of the original four—the one without tansy. The following 1743 American receipt wisely omitted the herb tansy.

> *Tansie very good*
> Take a pint of Milk as much Cream as much Spinage juice fifteen Eggs a little Salt leave out ½ the Whites strain them into the other things put them in near a pint of greated Bread or Bisket a whole Nutmeg greated as much Sugar as will make it very sweet thicken over the fire in a Skillet as thick as hasty Pudding then butter a Dish & bake it half an hour.
> —"Jane Randolph her Cookery Book" (1743), in Harbury, *Virginia's Cooking Dynasty*, 380–81

Tarragon. *Artemisia dracunculus*

Tarragon was commonly known as "the sallade herbe" and was "cherished in gardens." It was also widely employed in sauces. Tarragon vinegar seems to have been a particularly popular condiment to keep on hand for the table and as a seasoning ingredient in the kitchen. A variety of herbal vinegar receipts are worth following in any century.[79]

> Tarragon is hot and drie . . . and not to be eaten alone in sallades, but joyned with other herbes, as Lettuce, Purslain, and such like, that it may also temper the coldnes of them.
> —Gerard, *Herball*, 249

> Tarragon . . . hot and spicy: The Tops and young Shoots . . . never to be secluded from our Composition especially where there is much Lettuce.
> —Evelyn, *Acetaria*, 163

> Tarragon is a very hot, peppery herb. It is used in soup and salads. . . . Its young and tender tops only are used. It is eaten with beef-steaks in company with shalots. . . . It must be dried, like mint, for winter use.
> —Cobbett, *American Gardener*, 144

Teasel, Fuller's Teasel, Fuller's Thistle. *Dipsacus fullonum*

The European garden teasel has strong, hooked awns, or prickles. It was "sown, in Gardens or Fields for the Use of Cloth-workers" and used "to raise the nap when fulling woolen cloth." In contrast, "Wild Teasle is all Things like the former, but that the Prickles are small,

FIG. 36. Garden Teasel

John Gerard, *The Herball*, 1636

soft, and straight, not hooked or stiff" and therefore less useful for fulling cloth.[80]

> Fuller's Teasel . . . for raising the nap on wollen cloths, by means of the crooked awns upon the heads; which in the wild sort, are not sufficiently hooked. . . . The heads are collected in August, as soon as they begin to turn brown, and exposed daily to the sun, till they become perfectly dry.
> —M'Mahon, *American Gardener's Calendar*, 324

> The distilled Water of the Leaves . . . is often used by Women to preserve their Beauty, and to take away Redness and Inflamations, and all other Heat or Discolourings.
> —Culpeper, *English Physician*, 333

> No other use it hath than in dressing cloth; unless what the Superstition of some common People give it . . . as a Charm against Agues.[81]
> —Quincy, *English Dispensatory*, 127

Thyme.
Common Thyme. *Thymus vulgaris*
Mother of Thyme. *Thymus praecox*
Lemon Thyme. *Thymus × citriodorus*

In our extensive examination of period receipts thyme appears second only to parsley in importance in the kitchen and in the garden. The typical bundle of herbs tossed into a stewpot would almost certainly start with thyme and parsley and include others of the cook's favorite herbs. Thyme was used fresh, dried, or infused in vinegar, wine, or brandy.

> Thyme makes a neat border for cottage gardens.
> —Phillips, *History of Cultivated Vegetables*, 2: 334

> This is cultivated in the Kitchen-Gardens as a Soup Herb, and also for medicinal Use.
> —P. Miller, *Gardeners Dictionary*

> It smelleth comfortably. It hath much use, namely in all cold meats, it is good for Bees.
> —W. Lawson, *Country House-wives Garden*, 83

> The lemon-thyme is less pungent than the common garden thyme, but much more grateful; therefore, it is used as a seasoning for veal and other meats, where lemon-peel would be used, thus answering the purpose of two distinct spices.
> —Phillips, *History of Cultivated Vegetables*, 2: 334

> It frequently gives ease . . . when topically applied to a caries [decayed] tooth.
> —*Edinburgh Dispensatory*, 297

> [For nervous disorders] let his breakfast be mother of thyme tea, gathered in June . . . [o]r, the common garden thyme, if the former cannot be had.
> —Wesley, *Primitive Remedies*, 91–92

> Mother of thyme . . . is a small creeping plant. . . . Its taste, smell, and medical virtues are similar to those of thyme, but weaker.
> —*Edinburgh Dispensatory*, 283

> If you make a Vinegar of the Herb [mother of thyme], as Vinegar of Roses is made . . . and anoint the Head with it, it presently stops the Pains thereof.
> —Culpeper, *English Physician*, 337

Viola.

Common garden and woodland Violets. *Viola* species

Sweet Violet. *Viola odorata*

Wild Pansy, Viola, Heartsease. *Viola tricolor*

Sweet violets, which may be purple or white, have a wonderful fragrance that most common garden and woodland violets lack. Both common and sweet violet flowers are pleasant for a garnish, whether fresh or candied. Violet leaves were counted among mild spring greens and used with spinach and other herbs, raw or cooked. Each viola is a "choice flower of delight."[82]

> In the Spring, when they are in Flower, they cast forth a most agreeable Perfume . . . so that it renders such Places very delightful at that season.
>
> —P. Miller, *Gardeners Dictionary*

> Of [wild] Violets there are many, but they have no odor. . . . [Violets in the medical garden] are white and have the true odor. [Yellow violets were also listed among medical plants in North Carolina Moravian gardens.]
>
> —Fries, *Records of the Moravians*, 2: 572, 576

> ***A Violet Water***
>
> Infuse some violets in cold brandy. When these have lost their colour, take them out, and put in new ones. Repeat this till you are satisfied with your tincture. When you take the violets out, you must press them gently; then sweeten that brandy according to discretion; and, if you choose, you may add again a little orange-flowers for the sake of odour. [An early Carolinian might well have substituted the readily available peach blossoms for orange flowers.]
>
> —*Valuable Secrets*, 156

Violet water (above) illustrates an herbal infusion in alcohol—a cordial—while the syrup below is based on a standard aqueous infusion. The following delightful eighteenth-century method echoes a Gerard receipt from a century earlier that suggested: "Put more Violets into the same syrrup; thus do three or foure times, the oftner the better. . . . Some do put a little quantitie of the juyce of Lymons . . . which doth greatly encrease the beauty thereof, but nothing of the vertue."[83]

> ***Surrop of Violetts***
>
> Cutt of[f] the blew of the leaves [petals] then take as much Water as will cover them in a Silver—Tankard set it into a Skillet of hot water let them Infuse leishurely and when your leaves are—very white and the Water blew then strain it from it and put to every point [pint] a pound of loaf sugar then set the same Tankard into the Skillet again and let the Surrop Infuse as before tell the scum Rises which you must take of[f] the Surrop and take care you do not boyl it to hy for fear it should Candey.
>
> —Pre-1744 Virginia cookbook, 47

> [In cases of] suppression of urine . . . take a spoonful of the juice of lemon sweetened with syrup of violets.
>
> —Wesley, *Primitive Remedies*, 127

> A syrup made from [violet] infusion . . . proves an agreeable and useful laxative for children.
>
> —*Edinburgh Dispensatory*, 302

> There is likewise made of Violets and sugar certain plates called Sugar Violet or Violet tables [tablets] or Plate, which is most pleasant and wholesome, especially it comforteth the heart and the other inward parts.
>
> —Gerard, *Herball*, 853

FIG. 37. Purple Garden Violet

John Gerard, *The Herball*, 1633

FIG. 38. "Wilde Pansies," Viola, Heartsease

John Gerard, *The Herball*, 1633

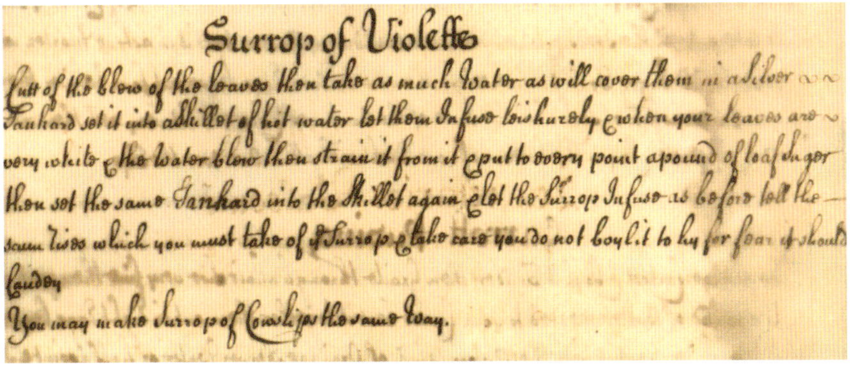

Surrop of Violetts

Cutt of the blow of the leaves then take as much Water as will cover them in a Silver Tankard set it into a Skillet of hot water let them Infuse leisurely & when your leaves are very white & the water blew then strain it from it & put to every point a pound of loaf Sugar then set the same Tankard into the Skillet again & let the Surrop Infuse as before till the same times which you must take of ye Surrop & take care you do not boyl it to high for fear it should Candey

You may make Surrop of Cowslips the same way.

FIG. 39. Surrop of Violetts

Pre-1744 Virginia cookery and cures manuscript

Caution: Do not confuse the inedible African violet houseplant (*Saintpaulia*) with the edible wild or garden violets.

Wormwood, Mugwort. *Artemisia vulgaris, Artemisia absinthium*

Roman Wormwood, Mugwort. *Artemisia pontica*

The generic name *Artemisia* covers a variety of herbs, several of which go by the common name "wormwood," being thought beneficial for expelling worms. Wormwood's use in preserving and flavoring ale may explain another name common to several of this confusing group—mugwort. Note that southernwood and tarragon are also of the genus *Artemisia*.

> Roman Wormwood, which is recommended in many Dispensaries to be used in Medicine. . . . The Plants however creep greatly by their Roots, so that when planted in a Garden, they multiply too fast, unless their Roots are every Year retrenched.
>
> —P. Miller, *Gardeners Dictionary*

> Wormwood is a strong bitter . . . against weakness of the stomach, and the like, in medicated wines and ales; but its use . . . is exceptional, on account of the ill relish and offensive smell. . . . It may be freed from these qualities partly by keeping, and totally by long coction.
>
> —*Edinburgh Dispensatory*, 115

> [To treat worm fever:] Drink, Beer brew'd pretty bitter, with Wormwood, instead of Hops.[84]
>
> —Tennent, *Every Man His Own Doctor*, 37

> Wormwood is an herb purely medicinal. . . . It must be dried and put by in bags for winter use.
>
> —Cobbett, *American Gardener*, 146

> *ye Eare Ach from worms*
>
> drop in . . . Juse of Wormwood Which kills them.
>
> —Allen, Medical memorandum

> It was a fashion with some people in this country [America] to drink . . . ale in which wormwood was infused. . . . To wormwood, as to every other bitter, has been ascribed an anthelmintic [deworming] virtue.
>
> —W. Barton, *Cullen's Treatise*, 2: 59

Green wormwood bruised is excellent for a fresh wound.
—Child, *American Frugal Housewife*, 36

A Worm Fever
Boil a handful of rue and wormwood in water; foment [apply outwardly] the belly with the decoction, and apply the boiled herbs as a poultice . . . and is likewise good if the fever be of the putrid kind.
—Thompson, Commonplace book, 145, from Wesley, *Primitive Remedies*, 70

[When stricken with a fever or pain in the head, neck, eye, or ear] apply the following Poultice to the Part where the pain lies: Boil the Leaves of Sage, Wormwood, and Rue together, and having beat them soft, grate Nutmeg thick upon them, and bind them on warm, renewing the same Night and Morning.
—Tennent, *Every Man His Own Doctor*, 13

Note: The above fomentation seems more likely to cause a rash than to rid a person of worms or calm a fever. Wormwood (especially when combined with rue as in the preceding and following quotes) is of questionable safety as a strewing herb on floors or among stored clothes. However, perhaps flea, mouse, or moth would indeed "scorn to meddle" where wormwood protects.

Where chamber is sweeped, and wormwood is strown,
No flea, for his life, dare abide to be known.
What savour is better, if physic be true,
For places infected, than wormwood and rue?
—Tusser, *Good Husbandry*, 172

Mix a little Wormwood . . . with your Ink, neither Rats nor Mice touch the Paper written with it. . . . This herb Wormwood being laid among Clothes, will make a Moth scorn to meddle with the Clothes as much as a Lion scorns to meddle with a Mouse, or an Eagle a Fly.
—Culpeper, *English Physician*, 362

Selected Wild Herbs from Farmyard, Field, and Forest

In Europe plants are not rejected because they are indigenous, on the contrary they are cultivated with due care; yet here [in America] we cultivate many foreign trifles, and neglect the profusion of beauties so bountifully bestowed upon us by the hand of nature.
—M'Mahon, *American Gardener's Calendar*, 72

A variety of wild herbs and shrubs supplemented the household larder. Some might have been invited into the garden; others were encouraged to flourish nearby. William Byrd echoed a widely held notion: "I believe it may pass for a Rule in Botanicks, that where any Vegetable is planted by the hand of Nature, it has more Vertue than in Places whereto it is transplanted by the Curiosity of Man." Among wild plants, those outlined here seem particularly interesting and useful.[85]

Dandelion, Piss-a-Beds. *Taraxacum officinale*
Dandelion is one of many wild herbs now common although not native to America. Dandelion, plantain, mullein, Queen Anne's lace, and numerous other familiar plants were brought here from Europe by accident or by design in the early days of settlement. Many other exotics have settled into this landscape too recently to have been part of the eighteenth-century experience.

FIG. 40. Dandelion

Elizabeth Blackwell, *A Curious Herbal*, 1739

The use of dandelion in treating urinary tract complaints is documented from Culpeper's 1770 England to Howard's 1836 American frontier:

> Dandelion, vulgarly called Piss-a-Beds . . . openeth the passages of the Urine . . . for which purpose the Decoction of the Roots or Leaves in White-wine, or the Leaves chopped as Pot herbs . . . and boiled in their Broth are very effectual. . . .
>
> You see here what Virtues this common Herb hath [reputed to cure nearly every complaint], and that's the Reason the French and Dutch so often eat them in the Spring: and now, if you look a little farther, you may see plainly, without a Pair of Spectacles, that Foreign Physicians are not so selfish as ours are, but more communicative of the Virtues of Plants to People.
>
> —Culpeper, *English Physician*, 111–12

In 1836 Ohio, dandelion's reputation as a diuretic continued.

> Producing an increased secretion of urine and serviceable in many nephritic affections. . . . If diuresis be desired more than this article alone will produce, the *Apium petroselinium* (common garden parsley) is a valuable adjunct.
>
> —Howard, *Botanical Medicine*, 261

> [Dandelion] is much commended in obstructions of the viscera, particularly of the liver.
>
> —Thomson, *Family Physician*, 3889

> This is a well-known and most wicked weed, in this country [America] as well as in England, and I am half afraid to speak of using it as food, lest I should encourage laziness. . . . In the spring . . . when nothing in the shape of greens was to be had for love or money, Dandelions were our resource; and I have always, since that time, looked at this weed with a more friendly eye.
>
> —Cobbett, *American Gardener*, 120

> Some have foolish Superstitions in gathering its Root, and wearing it as an Amulet against some Distempers of the Eyes; and also against Tertian Agues.
>
> —Quincy, *English Dispensatory*, 217

> The Root cures Diseases of the Eyes, and tertian Agues by way of Amulet, being cut into pieces, and hung about the Neck nine Days; others give the whole Root.
>
> —*The South Carolina and Georgia Almanack for the Year of Our Lord 1758*, Charleston: John Tobler, 1758

> The milky juice of the stem of this plant removes freckles from the skin.
>
> —Howard, *Botanical Medicine*, 262

Ground Artichoke, Jerusalem Artichoke, Potato of Canada. *Helianthus tuberosus*

In 1709 John Lawson reported: "Among Garden-Roots that thrive well in Carolina are . . . Ground Artichokes." More commonly known today as Jerusalem artichoke, this hardy American sunflower was also appreciated in the Old World. "The root spreads immoderately, multiplies very quickly, and is with difficulty cleared out of land where it is once planted." The edible tubers were widely enjoyed despite their reputation for causing flatulence.[86]

> Some boil them . . . and after stew them with sacke [a fortified white wine; we substitute sherry] and butter, adding a little ginger; others bake them in pies, putting Marrow, Dates, Ginger, Raisons of the Sun, Sacke, &c. . . . But in

FIG. 41. Jerusalem Artichoke

John Gerard, *The Herball*, 1633

my judgement . . . they stirre and cause a filthie loathsome stinking winde.

—Gerard, *Herball*, 754

The roots being the eatable part, are large fleshy tubers, much resembling a potatoe; are in perfection in autumn and all the winter, and are wholesome palatable food, when properly dressed.

—M'Mahon, *American Gardener's Calendar*, 206

Jerusalem artichokes are boiled and dressed in the various ways [as potatoes]. . . . They should be covered with thick melted butter, or a nice white or brown sauce.

—M. Randolph, *Virginia Housewife*, 120

The Potato's of Canada are by reason of their great increasing, growne to be so common here with us at London, that even the most vulgar begin to despise them, whereas when they were first received among us, they were dainties for a Queene.

—Parkinson, *Paradisi*, 518

[If] any one please to put them [roots] into boiling water, they will quickly become tender, so that, being peeled, sliced, and stewed with butter and a little wine, they will be as pleasant as the bottom of an artichoke. . . . The root nearly resembles the flavour of the artichoke bottom . . . [thus they were] improperly called Artichokes.

—Phillips, *History of Cultivated Vegetables*, 1: 296

They are very good roasted; they are then served in a napkin, and melted butter is eat with them. They are also mashed, and made in pie.

—Dods, *Cook and Housewife's Manual*, 208

They were formerly baked in pies . . . but the too frequent use, especially being so plentiful and cheap, hath . . . rather bred a loathing than a liking of them.

—Phillips, *History of Cultivated Vegetables*, 1: 296

The difficulty being, not to get it to grow, but to get the ground free of it, when once it has taken to growing. . . . The best way is to keep it out of the garden; and to dig up the corner of some field, and throw some seed or some roots into it.

—Cobbett, *American Gardener*, 127

[Jerusalem artichoke] yields most abunduntly in a good soil, and will thrive well even in poor land; affords good winter food, particularly for cows, sheep and hogs—little trouble attending the culture of this very valuable root.

—*North-Carolina Chronicle*, October 11, 1790

Lambsquarter, Lambs Quarter, Goosefoot, Wormseed. *Chenopodium albidum*

John Lawson listed lambsquarter among familiar plants he found growing in America in 1709. Philip Miller's *Gardeners Dictionary* described several species of European *Chenopodium*, both wild and cultivated. In the north of England it "was formerly cultivated in Kitchen Gardens . . . which in the Spring Season, they dress in the same Manner as Spinach." However, "in most parts of England" lambsquarter was "seldom cultivated, except in some Physick Gardens." Miller also described a native American species with an additional attribute: "The Seeds are called Worm Seed; I suppose from some Quality contained in it, which destroys Worms in the Body." *Chenopodium* seeds and leaves have served as basic subsistence plant foods since prehistoric times. Today's botanists recognize nutritional benefits of wild lambsquarter seeds similar to the closely related South American quinoa.[87]

Chenopodium . . . used as kale. Only the young plants a few inches in height were used. [Wild kalelike spring greens included lambsquarter *Chenopodium*, sorrel *Rumex*, and poke *Phytolaca*]. . . . Vinegar is placed in a special container on the table to be used on the [wild] kale. I must confess that this dish tastes very good.

—Kalm, *Travels*, 1: 285

Monarda, Bergamot, Bee Balm, Oswego Tea. *Monarda didyma*
Horse Mint. *Monarda punctata*, *Monarda fistulosa*

Settlers discovered this colorful relative of the mints growing in frontier regions. Monarda was among the curious herbs they sent back to their European homelands.

> [Monarda] grows naturally in North America, where the Inhabitants frequently use the Leaves for Tea, so it is commonly called Oswego Tea, by which Title it was brought to England. . . . This is a very ornamental Plant in Gardens, and the Scent of the Leaves is very refreshing and agreeable to most People.
>
> —P. Miller, *Gardeners Dictionary*

> A tea made either of the green or dried leaves [of Horse Mint], will stop vomiting. . . . [Also] bringing on the menses, or courses of women. . . . It may be placed on a footing with Rosemary, Pennyroyal, and many other simple herbs. All this, however is well known to every old lady in the country.
>
> —Gunn, *Domestic Medicine*, 400

The following simple remedy from early-nineteenth-century Tennessee was attributed to "An African Negroe." Thomas Johnson published a variation of it in *Every Man His Own Doctor*; another was found on an undated scrap among "miscellaneous cures" in the Cameron family's papers.[88]

> Cure for the gravel [kidney stone]—take red onniuns and horse mint and beat them together in a morter and put a little water to it and Sqeese out the juice and Drink it . . . the Strength of two onions and a Handful of mint must be taken in a day.
>
> —Thompson, Commonplace book, 36

> I have observed that [hummingbirds] have fluttered chiefly about . . . the Monarda with crimson flowers. An inhabitant of the country is sure to have a number of these beautiful and agreeable little birds before his window all the summer long, if he takes care to plant a bed with all sorts of fine flowers under them.
>
> —Kalm, *Travels*, 1: 112

Mullein, Hig Taper, Hag Taper, Torches, Cows-Lungwort. *Verbascum thapsus*

Mullein, a European native, had a variety of medicinal uses, primarily for chest complaints and piles (hemorrhoids). The smoke from burning mullein was a suggested treatment for piles; more mainstream treatments involved the leaves and flowers as illustrated in the following chronology.

> *[1742 England:] For the Piles*
>
> Take of the tops of parsley, of mullet [mullein], and of elder-buds, of each one handful; boil in a sufficient quantity of fresh butter till it looks green, *and* has extracted the smell of the herbs; strain, and anoint the place with it three times or four times a day.
>
> —Smith, *Compleat Housewife*, 321

FIG. 42. Mullein

Elizabeth Blackwell, *A Curious Herbal*, 1739

[1770 England:] And the Oil made by the often Infusion of the [mullein] Flowers, is of very good Effect for the Piles.

—Culpeper, *English Physician*, 225

[1738 South Carolina:] A little fine Treacle spread upon a leaf of Mullein & laid to the Piles or Hemorhoids cureth the same; an ointment made of the Leaves & old Hogs Grease has the same effect.

—Unknown author, Herbal and household book, from Gerard, *Herball*, 774

[To prevent hemorrhoids:] Boil a Handful of Mullein Leaves in a Pint of new Milk, and sweeten it with Syrup of Violets. Drink this every Night, for 6 Weeks.

—Tennent, *Every Man His Own Doctor*, 54

Cure for Lax [Diarrhea]
Take of the sort of Mullen that doth not go to seed [first-year growth] . . . top leaves with each bud . . . & boile them in a pint of new milk . . . & drink it.

—Lenoir, Medical memorandum, 17

The [mullein] root boiled in Water & drank is good for them that are broken & hurt inwardly & prevaileth much against the Old Cough.

—Unknown author, Herbal and household book, from Gerard, *Herball*, 774

A strong [mullein] tea, taken internally, is good for agues, croup, asthma, cough, bleeding at the lungs; and externally, in wash, for piles, scalds, and burns. The flowers are said to be better than the leaves.

—Howard, *Botanical Medicine*, 297

The Countrey people . . . doe give their cattell the leaves to drinke against the cough of the lungs.

—Unknown author, Herbal and household book, from Gerard, *Herball*, 774

Mullein was collected from the wild or invited into the flower garden. The fuzziness of the leaves recommended them for uses other than medical.

May be admitted in large Gardens, for the Variety of their *hoary* Leaves, together with the extreme Sweetness of their Flowers, which have a Scent somewhat like Violets.

—P. Miller, *Gardeners Dictionary*

[Mullein] leaves being rough and bearded . . . are used instead of cotton [wick] in a lamp.

—*Treatise on Gardening*, 38

The report goeth . . . that figges do not putrifie at all that are wrapped in the leaves of Mullein.

—Gerard, *Herball*, 774

Mustards. Various members of the Brassicaceae family, wild and cultivated.
White. *Brassica hirta*, *Sinapis alba*
Black. *Brassica nigra*
Rocket, Winter Cress. *Barbarea vulgaris*, *B. verna*
Numerous Wild Mustards. *Capsella bursa-pastoris*, *Lepidium campestre*, and others

The sharp taste of young mustards recommends them for salads. At the dawn of the eighteenth century, John Evelyn wrote that mustard is "so necessary an Ingredient to all cold and raw Salleting, that it is very rarely, if at all to be left out." Mustard "is sometimes found wild, but for culinary and medical uses is cultivated in gardens." Winter cress, among other wild and naturalized mustards, may be gathered from local fields, or cultivated in your garden if a greater quantity is desired.[89]

> In winter, when salad herbes bee scarce, [winter cress] is thought to be equall with Cresses of the garden, or Rocket.
>
> —Gerard, *Herball*, 243

> The young and green mustard plants which are so readily and easily reared in the spring, are perhaps the most beneficial, as well as the most agreeable addition to our salads.
>
> —Phillips, *History of Cultivated Vegetables*, 1: 383

> Cress is very good in salads along with lettuces, white mustard, or rape . . . cut before it comes into rough leaf.
>
> —Cobbett, *American Gardener*, 118

The white mustard was suggested for salad greens, while black or brown mustard seeds were collected for the condiment. Details for growing mustard outlined in the *Records of the Moravians in North Carolina* document an early interest in producing quantities of this popular herb for home use and for market.

> The seeds are pounded in a mortar; or bruis'd with a polished Cannon-Bullet, in a large wooden Bowl-Dish.
>
> —Evelyn, *Acetaria*, 174

> [Mustard sauce] is prepared in a variety of ways; plain, with broth or boiling water; mild with milk or cream, or with the addition of a little sugar; pungent, with water in which garlic, horseradish, &c. is boiled; it is also prepared with the flavoured vinegars, with cayenne . . . and even with spirits.
>
> —Dods, *Cook and Housewife's Manual*, 258

Lettice Bryan's *Kentucky Housewife* agreed with each of Dods' variations on mustard and added a few refinements. Bryan specified tarragon vinegar, recommended shallots rather than garlic, and added the following receipt:

> ***Made Mustard***
>
> Grind and sift three pounds of black mustard seed; mix with it one pound of salt, and liquefy it sufficiently with currant wine, adding four spoonfuls of sugar to each pint.
>
> —Bryan, *Kentucky Housewife*, 162

The 1736 *English Dispensatory* listed brown-seeded mustards as "much more used in Sauce, than in Medicine: altho' they are . . . excellent in . . . the Decays and Defluxions of Old Age." Medical usage was evident not only in England but persisted in early America, as noted by William Lenoir in western North Carolina.

> ***[For Rheumatism]***
>
> Take half pint of Mustard seed beaten to powder & put in a Gallon of wine & give a Common Dram Glassful 3 times a day . . . Cyder answers as well as wine.
>
> —Lenoir, Medical memorandum, 4

> ***Cure for Poison***
>
> Take 2 Tea spoonfulls of made Mustard, mixed in warm water. It acts as an instantaneous emetic.
>
> —Lenoir, Medical memorandum, 56

> The Seed bruised, mixed with Honey and applied . . . taketh away the Marks and black and blue Spots of Bruises . . . the Roughness or Scabbiness of the Skin, as also the Leprosy, and Lousy Evil. It helpeth also the Crick in the Neck.
>
> —Culpeper, *English Physician*, 228

> ***Chapped Hands (To Prevent)***
>
> Wash them with flour [flower] of mustard.
>
> —Wesley, *Primitive Remedies*, 43

Plantain.
Broadleaf, Common Plantain. *Plantago major*
Ribwort, Narrowleaf Plantain. *Plantago lanceolata*

William Lawson found "all sorts of Plantain" growing in the Carolinas in 1701. According to Swedish explorer Peter Kalm (c. 1750) Native Americans gave plantain "a name which signified the [Englishman's] foot, for they say that wherever a European had walked, this plant grew in his footsteps." Moravian settlers noted, "Plantain exists in sufficient quantities. . . . In the spring the leaves can be cooked like cabbage." The leaves were also employed as a dressing for wounds and sores, as seen in the following quotations from 1770 and 1802.[90]

> Plantains are singular good Wound Herbs to heal fresh or old Wounds or Sores, either inward or outward.
>
> —Culpeper, *English Physician*, 259

> The leaves are now commonly applied to fresh wounds and cutaneous sores.
>
> —Thomson, *Family Physician*, 413

According to Joseph Doddridge, on the Virginia frontier "the white plantain held a high rank" for curing snakebite. The well-known plantain and horehound dose for poisoning concocted by the enslaved African Caesar was particularly recommended against animal poisons, including snakebite and the bite of a rabid animal. In North Carolina, William Lenoir wrote in his medical memorandum book of a "notable and respectable Physician . . . [who] for forty years . . . had never known a single instance of [Caesar's Cure] failing to afford Relief to persons bitten by Snakes." Plantain, with or without horehound, was widely endorsed for counteracting poisons. For "A Venomous Sting" John Wesley advised "a poultice of bruised plantain and honey."[91]

> Some have had strange Notions of [plantain] resisting Poisons. . . . [Ancient legend claimed] that a Toad before she engages with a Spider, will fortify herself with some of this Plant; and that if she come off wounded, she cures herself afterwards with it.
>
> —Quincy, *English Dispensatory*, 95

Pokeweed. *Phytolacca americana*

Caution: The entire pokeweed plant contains toxins that become more concentrated as the plant matures. Nevertheless, the southern tradition of eating a "mess" of poke greens or "poke sallet" in spring continues among rural folk and those exploring wild edibles. To avoid gastrointestinal distress, do not eat poke shoots and leaves raw. Pokeweed should be gathered only in early stages before flower buds appear and stalks become tinged with red. Young shoots or leaves must be boiled and drained once before proceeding to prepare as other vegetables.

> The Inhabitants of North America boil the young Shoots of this Plant, and eat it like Spinach. The Juice of the Berries stain Paper and Linen of a beautiful purple Colour, but it will not last long. If there could be a Method of fixing the Dye, it might be very useful.
>
> —P. Miller, *Gardeners Dictionary*

> ***[Poke Greens or Poke Sallet]***
>
> Vegetables not being over abundant in these back woods at any time, and there being a great deficiency of them in the spring of year, we adopt the custom of the inhabitants

FIG. 43. Broadleaf Plantain

Maria Sibylla Merian, *Erucarum Ortus*, 1718

who gather the leaves of the poke-plant, just as they shoot above ground and are tender and soft; it is no bad substitute for spinnage, and greatly resembles it in flavor, yet great care must be taken in gathering of it, that it is neither too old, nor the stalk grown.

—Anburey, *Travels*, 2: 218

Young shoots, when boiled, may be eaten with impunity, and are justly deemed a great delicacy. By many, they are preferred to the finest asparagus.

—W. Barton, *Collections*, 2: 27

Poke Asparagus

Poke, when gathered at a proper season and nicely dressed, is considered by many people to be as fine as asparagus. It is the young tender stalks of the common pokeberry plant that shoots forth in the spring, and should be gathered when about the size of asparagus of the largest size. Scrape and cut the stalks of equal lengths, soak them in fresh water for an hour or two, and then boil them as directed for asparagus. Serve them warm, lay round them on the edge of the dish some small buttered toasts, and send with them a boat of melted butter, seasoned with pepper.

—Bryan, *Kentucky Housewife*, 204

Bryan recorded a method for seasoning poke with cured pork that will be thoroughly familiar to today's traditional southern cooks:

Poke Tops

Poke tops, or sprouts that put forth in the spring of the year, are considered fine sallad by many people. Gather them when very young and tender, pick them carefully, pour boiling water on, and let them stand for an hour or two, to draw out the strong taste. Having ready a pot of boiling water, in which a piece of bacon has been boiled till nearly done, put the poke into it, and let it boil with the meat till it is tender. It will take but a short time to boil it sufficiently, as it is not good when boiled very soft. Serve it with the meat, drain it well, and have salt, pepper and vinegar to season it at table.

—Bryan, *Kentucky Housewife*, 204

Note that the above authors, spanning nearly a century, stressed that only very young leaves and shoots may be eaten, and only after boiling. The initial cooking or soaking water should be discarded.

Pokeberries should never be eaten. Children are particularly at risk because the berries appear quite tempting. Nevertheless, *The Carolina Housewife* suggested pokeberry juice for coloring ice cream, and in Benjamin Barton's 1810 *Collections . . . towards a Materia Medica* we read: "The ripe berries, infused in brandy or wine . . . are a popular remedy for rheumatism, in many parts of the United States."[92]

Poke-root, boiled in water and mixed with a good quantity of molasses, set about the kitchen, the pantry, &c. in large deep plates, will kill cockroaches in great numbers, and finally rid the house of them. The Indians say that poke-root boiled into a soft poultice is the cure for the bite of a snake.

—Child, *American Frugal Housewife*, 19

Cure for the Bite of a Poisonous Snake

Boil Poke Root till soft and apply it to the wound.

—*Gale's North Carolina Almanac for 1830*, North Carolina State Archives, Raleigh, PC 89.11

FIG. 44. Narrow-Leaved Plantain, Ribwort

Elizabeth Blackwell, *A Curious Herbal*, 1739

Purslane. *Portulaca oleracea* and other *Portulaca* species

There seem to have been broad differences of opinion regarding purslane—whether it "is fitly sowne in gardens" or "one of the worst weeds." Nevertheless, purslane was recommended for a variety of culinary uses. In addition to salads, note elsewhere in this volume purslane's use in soup with chervil, with spinach to make a potato puree green, and in herb sauce.

> [Sow] in Aprill, in the alleyes [alleys] of the Garden betweene the beds . . . After [Cucumbers, Melons, and Pumpkins] have been taken away they have sowen Purslane.
>
> —Parkinson, *Paradisi*, 499

> [Garden purslane] is fitly sowne in gardens, and in the waies and allies thereof being digged and dunged; it delighteth to grow in a fruitfull and fat soile not dry. [Wild purslane] commeth up of his owne accord.
>
> —Gerard, *Herball*, 521

> Purslain wild, and that of a larger Size which grows in the Gardens.
>
> —J. Lawson, *New Voyage*, 83

> In gardens it was one of the worst weeds.
>
> —Kalm, *Travels*, 1: 355

> A mischievous weed that Frenchmen and pigs eat when they can get nothing else. Both use it in salad, that is to say raw.
>
> —Cobbett, *American Gardener*, 140

> Generally entertain'd in all our Sallets mingled with the hotter Herbs: 'Tis likewise familiarly eaten alone with Oyl and Vinegar.
>
> —Evelyn, *Acetaria*, 158

> Purslane is pickled as the cowcomber and you may serve them together.
>
> —La Varenne, *French Cook*, 203

> Among the pot herbs, tarragon and purslane can also be used in green salads . . . with milder herbs.
>
> —Von Rumohr, *Essence of Cookery*, 164

Woodsorrel.

See **Sorrel**.

Yarrow, Milfoil, Woundwort, Nosebleed.

Achillea millefolium

John Lawson wrote of an "abundance" of yarrow among useful herbs he found growing in the Carolinas. Note yarrow's contradictory medical uses suggested by the common names "woundwort" and "nosebleed." One wonders how the herb came to be thought to cause the nose to bleed yet also stanch bleeding from wounds, as seen in the proposed treatments below:[93]

> The leaves being put into the nose, do cause it to bleed, and ease the paine of the megrim [migraine].
>
> —Gerard, *Herball*, 1073

> Serviceable in all Kinds of Haemorrhages.
>
> —P. Miller, *Gardeners Dictionary*

> ***Bleeding Piles***
>
> Live all the Time upon a cooling Diet, without Meat; and only drink Burnet or Yarrow Tea.
>
> —Tennent, *Every Man His Own Doctor*, 44

FIG. 45. Purslane

John Gerard, *The Herball*, 1636

Wounds. If you have not an honest surgeon at hand Apply juice or powder of yarrow.
—Wesley, *Primitive Remedies*, 132

The flowers of milfoil are considerably stronger in aromatic flavour than the leaves.
—*Edinburgh Dispensatory*, 231

It stays the shedding of Hair, the Head being bathed with the Decoction of it.
—Culpeper, *English Physician*, 365

[Yarrow] makes a good breast tea [for chest complaints]. Some eat it in the morning with bread and butter.
—Fries, *Records of the Moravians*, 2: 571

Of Several Useful Trees and Shrubs

Crab Apple. *Malus* species

The pleasing acidity of the crab apple, native to America, satisfied the settlers' craving for tart flavors. The following quotation indicates the priority placed on these native fruits.

One or more of these indigenous apple trees ought to be planted in every orchard, in honor of their native tenancy of our forests, as well as for the convenience of our ladies, who are very fond of them for preserves.
—Doddridge, *Notes on the Settlement*, 73

Crabapple . . . in the woods of Virginia, the Carolinas, and Georgia . . . [bears tiny] apples of a very sour and disgusting taste. This fruit is collected and preserved candied with sugar; and sometimes from it is distilled a kind of spirit called crabapple brandy.
—Castiglioni, *Viaggio*, 433

Crab-Apple Shrub
Gather your crab-apples when ripe, break them up with a wooden mallet, and boil them for a few minutes in just enough water to cover them: then strain the liquid and set it by till next day; mix with it a pound and a half of loaf sugar to each quart of the liquid, and boil and skim it well. When it is cold, put it up in quart bottles, putting in each a gill [one-half cup] of brandy, and mix it with iced water for a summer's drink.
—Bryan, *Kentucky Housewife*, 394

Verjuice is the fermented juice of various sour fruits such as crab apples and unripe grapes. Verjuice was sometimes substituted for vinegar, wine, or juice of the expensive and less readily available lemon to adjust the balance of sweetness or tartness of a dish.

[Crab apples] are so extremely sour . . . as not to be eaten raw; they are therefore used only for their expressed Juice, which serves in some Countries instead of Vinegar.
—Quincy, *English Dispensatory*, 219

Verjuice
I have been used to stamp and press out the juice of crabs [crab apples] . . . which I kept in a cask by itself in my cellar. . . . Crab-juice is so necessary for abundance of uses . . . that none ought to be without it.
—Ellis, *Country Housewife's Family Companion*, 314

Dogwood. *Cornus florida*

Dogwood was considered medicinal because of its bitter bark, which suggested dogwood as a substitute for the costly Peruvian bark (cinchona) widely used in treating the intermittent fevers of malaria. The South

American bark contains quinine, a specific for curing malaria. Although dogwood did not possess quinine, this native tree was found in most forests and therefore could be gathered fresh at no cost when needed.

> [Dogwood bark] was used advantageously in America for intermittant fevers.
>
> —Castiglioni, *Viaggio*, 372

> The bark of this American tree is an excellent substitute for the Peruvian bark, especially as a remedy for intermittents.
>
> —W. Barton, *Cullen's Treatise*, 2: 82

> I considered [dogwood bark] greatly superior, not only on account of our always being able to procure it fresh from the tree, but because the Peruvian bark is old before it reaches this country.
>
> —Gunn, *Domestic Medicine*, 366

> For want of such necessary implements as toothbrushes, we substitute this wood.
>
> —Anburey, *Travels*, 450

> These are tooth brushes of nature's presenting, and are infinitely better than those made of hog's bristles, and filled with snuff, and such other delightful aromatics!
>
> —Gunn, *Domestic Medicine*, 367

> ***Cure for the Collic***
>
> [Take] three dogwood buds or buttons that grows like the bud, and beat them and put it in a spoonful of water and take it.
>
> —Thompson, Commonplace book, 13

> Dog's Wood is so called because it stinks. . . . [It] does not become large, but is hard, and is good on the turning-lathe.
>
> —*Records of the Moravians*, 2: 561

Elder. *Sambucus canadensis* (American); *Sambucus nigra* (European)

Wine made from the elder berries was considered exceptional; a pleasant wine was also made with the flowers. Children used the stems for popguns. Elderblow (elder blossoms) infused in brandy or a distilled aqueous solution may be used in the place of rose, orange, or peach blossom flavoring agents. A historically approved condiment results from infusing elder blossoms in vinegar.

> I hold it needless to write any Description of this . . . every Boy that plays with a Pot-gun [*sic*] will not mistake another Tree instead of Elder.
>
> —Culpeper, *English Physician*, 121

> Elder grows in sunny spots in the Bottoms and along the creeks.
>
> —Fries, *Records of the Moravians*, 2: 562

> Elder-wine, made of the elder-berries, is a rich and expensive preparation [because of the quantity of sugar required]. . . . Elder-wine is the pride of many English housewives, and no expense nor pains are spared in its preparation.
>
> —Dods, *Cook and Housewife's Manual*, 469–70

> The flowers . . . have an agreeable aromatic flavor, which they give over in distillation with water, and impart by infusion to vinous and spirituous liquors.
>
> —*Edinburgh Dispensatory*, 274

> Gather Elder-Flowers when they are dry, and pick them from the Stalks; let them dry in the Shade, and then put an Ounce to each Quart of White-Wine Vinegar,

FIG. 46. Elder

Elizabeth Blackwell, *A Curious Herbal*, 1739

to stand in the Vinegar for two Months, then pour the Vinegar from them for use.

—Bradley, *Country Housewife*, 1: 96

Elder-blow tea . . . is cool and soothing, and peculiarly efficacious either for babes or grown people, when the digestive powers are out of order.

—Child, *American Frugal Housewife*, 37

Gather . . . Elder Flowers, just as they are beginning to open, for that is the Time of their Perfection, they have just then a very fine Smell and a spirited Taste, but afterwards they grow dead and faint . . . having a sickly Smell.

—Cleland, *New and Easy Method of Cookery*, 209

Some people add elderflowers, which are known to be sudorific [promote sweating], to sweet dishes. [When past their prime] they impart a rather flat, sweet flavour which is inclined to become unpleasant.

—Von Rumohr, *Essence of Cookery*, 163

The spring Buds and tender Leaves, excellently wholsome in Pottage at that Season of the Year.

—Evelyn, *Acetaria*, 148

Elderblow was also made into fritters. "Each Fritter is one of the Bunches of Elder, with as much Batter as agreeably covers it." Alternatively, a quantity of individual elder blossoms may be stirred into a fritter batter as simple as milk and flour, or Cleland's more elegant receipt of flour, eggs, white wine, salt, and cinnamon. Elder flowers might be substituted, as in the herb fritter receipt under **Clary**.[94]

Tradition suggested a range of medical uses for elder. Nicholas Culpeper's *English Physician* asserted: "Leaves or Bark of Elder stripped upwards as you gather it, causeth Vomiting," supposedly by acting upwardly on the body in an example of the ancient Doctrine of Signatures. Similarly, "stripped downwards, it purgeth downwards."[95]

The inner [elder] bark, flowers, and berries, are used as medicine, being cathartic, emetic, and diueretic.

—Howard, *Botanical Medicine*, 290

For a gentle purge [t]ake Elder Roots peach Blossoms & Rose leaves Stew or Boyl them.

—Horry, Household book

A cure for a white swelling of any description

Take Elderflowers . . . Boyl them in new milk; and make a poultice with the crums of light wheet bread . . . and bind it to the sore as warm as can be born[e].

—Thompson, Commonplace book, 138

[To treat jaundice:] Decoction of the inner Bark of Elder, and the Root of Sassafras.

—Tennent, *Every Man His Own Doctor*, 39–40

Persimmon. *Diospyros virginiana*

The sweet fruit of this native tree must have been a welcomed addition to simple backcountry fare. When exploring the Carolina backcountry, John Lawson penned a colorful description of the persimmon: "These Fruit, when ripe, is nearest our Medlar; if eaten before, draws your Mouth up like a Purse. . . . The Fruit is rotten, when ripe, and commonly contains four flat Kernels, call'd Stones."[96]

With regard to fruit trees there is little choice. The apple, peach and pear are common; the persimmon deserves culture.

—*The North-Carolina Chronicle; or Fayetteville Gazette*, September 27, 1790

The Taste of [persimmons] is so very rough, it is not to be endured, till they are fully ripe, and then they are a pleasant Fruit. Of these some Vertuosi make an agreeable kind of Beer; to which Purpose they dry them in Cakes, and lay them up for Use.

—Beverly, *History of Virginia*, 2: 12

Mr. Bartram, related that [persimmons] were commonly put upon the table amongst the sweetmeats and that some people make a tolerably good wine of them. . . . I must own that those who praised this fruit as an agreeable one have but done it justice. It really deserves a place among the most palatable fruits of this land, when the frost has entirely removed its bitterness.

—Kalm, *Travels*, 1: 69

Where apples and cyder are scarce, the inhabitants gather the persimmons, after they are perfectly ripe, knead them into a kind of dough, or paste, with wheat bran, which they form into loaves, and bake in ovens; of these they brew a fermented liquor, which is called persimmon beer. This serves for their common drink, and is tolerably pleasant and wholesome. It is sometimes, though rarely, distilled into brandy.

—Smyth, *Tour*, 151

The barks of the Common Sassafras and Persimmon have likewise been found useful in intermittents [malaria or fevers in general].

—B. Barton, *Collections*, 1: 11

John Lawson described Native Americans' use of persimmon seeds in gaming.

Plays and Games: as, with the Kernels or Stones of Persimmons, which are in effect the same as our Dice, because Winning or Losing depend on which side appear uppermost, and how they happen to fall together.

—J. Lawson, *New Voyage*, 180

Redbud, Judas Tree, Sallad Tree.

Cercis canadensis

In the Carolinas as in the mid-Atlantic colonies *Cercis* was known as "the sallad tree." The blossoms do indeed have a pleasant, mildly tart flavor and supply a beautiful fresh touch to a salad.[97]

The Red-Bud-Tree . . . is the best Sallad, of any Flower I ever saw.

—J. Lawson, *New Voyage*, 106

American Judas tree, is one of our most beautiful early flowering and ornamental plants; and may be propagated, by sowing its seeds in March.

—M'Mahon, *American Gardener's Calendar*, 268

Sassafras, Ague Tree. *Sassafras albidum*

Gerard's 1633 *Herball* noted: "The Spaniards and French men have named this tree, Sassafras . . . for want of an English name we are contented to call it the Ague tree, of his vertue in healing the Ague." Sassafras retained a reputation for healing fevers and purifying the blood through at least three centuries. Be aware, however, that sassafras contains safrole, a weak carcinogen found also in nutmeg, cinnamon, and pepper.[98]

All has an odor, leaves, wood and blossoms; and it is much used in medicine to purify the blood.

—Fries, *Records of the Moravians*, 2: 562

The root, bark, or flowers, made into a tea, is used considerably by the people in the country.

—Gunn, *Domestic Medicine*, 362

It bears a white [pale yellow] Flower, which is very cleansing to the Blood, being eaten in the Spring, with other Sallating. . . . The Bark of the Root is a Specifick to those afflicted with the Gripes.

—J. Lawson, *New Voyage*, 101

Cure for Sore Eyes from** Salem Almanack **advertisement, 1829
Take small Sticks of Sassafras split into four pieces put them in a Vessel with cold Spring water, they impart Glutinous matter to the water. Wash the sore Eyes with it; and it cures them without smart or heat.

—Lenoir, Medical memorandum, 54

[For consumption] chew Sassafras Root every Morning fasting. . . . [Heartburn] is remov'd by chewing Sassfras Bark, or by a Decoction of it. . . . [If] you wou'd draw a Swelling to a Head, chew Sassafras Root, and apply it by Way of Poultis.

—Tennent, *Every Man His Own Doctor*, 17, 26, 67

[Drink] half a pint of Sassafrass Tea after every dose of [strong laxative].

—"Jane Randolph her Cookery Book" (1743), in Harbury, *Virginia's Cooking Dynasty*, 304

[For] a depression of the animal spirits . . .
The occasionally taking such medicine as Salts &c. and a moderate use of Sassafras Tea . . . [also] a sufficiency of moderate exercise . . . [and] a plentiful supply of what we generally stile fun.

—Adam Springs to Polly Springs, letter, February 5, 1798[99]

Watery infusions of sassafras . . . are commonly drunk as tea; but the spirituous tincture, or extract, which contains both the volatile and fixed parts of the medicine, appears to be preferable.

—Thomson, *Family Physician*, 420

[For dropsy (edema):] An Ancient Gentleman, who trusted to the Remedies of our own Country, with . . . happy Effect. He drank the Decoction of Sassafras, as soon as he got up, and chew'd the Root of it all the rest of the Morning, 'till Dinner.

—Tennent, *Every Man His Own Doctor*, 42

The spicy scent of this small native tree, particularly the roots, captured the attention of the first settlers. Sassafras was one of the first New World exports.

This is somewhat come into fashion in Families, by way of Tea; which the Shavings of it make agreeably enough; but the Scandal of its being good in Veneral Cases is a great detriment to its Credit.

—Quincy, *English Dispensatory*, 105

Some Englishmen [in mid-eighteenth-century Pennsylvania] related, that some years ago it had been customary in London, to drink a kind of tea of the flowers of sassafras, because it was looked upon as very healthful, but upon recollecting that the same potion was much used against the venereal disease, it was soon left off, lest those who used it should be looked upon as infected with that disease.

—Kalm, *Travels*, 1: 180

FIG. 47. Sassafras with Small Green-Crested Flycatcher

John James Audubon, *Birds of America*, vol. 1, 1840

Although first acclaimed for its supposed medicinal value, sassafras had other virtues as well.

> The bark of this tree [sassafras] is used . . . in dying worsted a fine lasting orange color, which does not fade in the sun. They use urine [as a mordant] instead of alum in dyeing, and boil the dye in a brass boiler, because in an iron vessel it does not yield so fine a color.
>
> —Kalm, *Travels*, 1: 78

> The wood and leaves give off a very pleasant fragrance, so that, some people believe it is excellent for keeping moths and other insects away from beds and wardrobes.
>
> —Castiglioni, *Viaggio*, 399

South Carolinian Sara Rutledge recorded a recipe for a soup of obvious Native American origin. This dish utilizes the pleasant flavor and mucilaginous property of sassafras leaves. An option to substitute pine instead of sassafras leaves seems less appealing.

> ***Seminole Soup***
>
> Take a squirrel, cut it up and put it on to boil. When the soup is nearly done add to it one pint of picked hickory-nuts and a spoonful of parched and powdered sassafras leaves—or the tender top of a pine tree, which gives a very aromatic flavor to the soup.
>
> —Rutledge, *Carolina Housewife*, 46

Rutledge collected another exotic soup flavored and thickened with sassafras leaves. Her gumbo merged English, French, African, and Native American tastes and ingredients. In this receipt the sassafras leaves seem intended more as a thickening agent than a seasoning. In soups and gumbos okra also serves as both a thickening agent and a flavoring. Peter Kalm reported that okra was "boiled in soups" and "reckoned a dainty by some people and especially by the negroes."[100]

> ***New Orleans Gumbo***
>
> Take a turkey or fowl, cut it up, with a piece of fresh beef; put them in a pot, with a little lard, an onion, and water sufficient to cook the meat. After they have become soft, add a hundred oysters, with their liquor. Season to your taste; and just before taking up the soup, stir in until it becomes mucilaginous two spoonsful of pulverized sassafras-leaves.
>
> —Rutledge, *Carolina Housewife*, 43

The 1839 *Kentucky Housewife* included the following receipt. Although ice creams were atypical for this historical period and were likely to be prepared only in the most elegant families, this method for sassafras ice cream was just too interesting to leave out:

> ***Sassafras Cream***
>
> May be made in every respect as directed for vanilla cream, and is very much liked. For this purpose take the red sassafras, which will color as well as flavor it:
>
> Boil [sassafras root chips] in a quart of rich sweet milk till the milk is sufficiently flavored; then take out the [chips]; and set the milk by to cool; after which stir into it the beaten yolks of half a dozen eggs, and simmer it again till a little thick, stirring it all the time. Add three quarters of a pound of sugar [one and one-half cups], and freeze it.
>
> —Bryan, *Kentucky Housewife*, 343–44

Sumac, Sumach, *Rhus glabra*

The late-eighteenth-century *Edinburgh Dispensatory* described multiple uses for sumac: "This tree, or shrub,

FIG. 48. Smooth Sumac

Mark Catesby, *Natural History of Carolina, Florida, and the Bahama Islands*, vol. 2, c. 1771

is cultivated in some places on account of the culinary uses of its fruits, and for the purposes of the dyers, &c. among us (in Britain) it is met with only in the gardens of the curious." The various parts of the sumac plant produce a surprising range of colors.[101]

> The water wherein the [Sumach] leaves hath bene boyled . . . maketh the heare [hair] blacke. . . . The seede of Sumach eaten in sauces with meate doth also stop all fluxes [diarrhea].
>
> —Dodoens, *Niewe Herball*, 692

> [Sumac] The bark, leaves, or berries, may be used as medicine, and possess valuable properties, being astringent, tonic, and diuretic. . . . The berries made into a tea and sweetened make a pleasant drink for children. . . . The berries and leaves are found equal to nutgalls in dying or making ink, giving a deep and permanent black.
>
> —Howard, *Botanical Medicine*, 289

Papaw's (or Papa's) Remedy for venereal diseases and yaws (a nonvenereal form of syphilis) appeared in contrast to the ancient herbal tradition of sumac as a cure for diarrhea. A strong decoction of the combined barks of sumac root, pine, and Spanish oak was recommended to induce "a strong vomit," and as a gargle and to bathe venereal sores. In 1729 a Virginia slave, James Papaw, received his freedom for sharing his celebrated method. Papa's Remedy was widely published in newspapers, in Tennent's *Every Man His Own Doctor*, and in the central North Carolina commonplace book of Rachel Allen. Nightly pills of deer dung and turpentine were recommended along with that emetic decoction.

> The wood is a well-known, good, black dye. Ink can be made from it.
>
> —*Records of the Moravians*, 2: 563

> [Sumach] fruit or berries are red. They are made use of for dyeing and afford a color like their own. . . . The branches boiled with the berries yield a black ink-like tincture. The boys eat the berries . . . but they are very sour.
>
> —Kalm, *Travels*, 1: 42

Sweet Shrub, Bubby Flower, Carolina All-spice. *Calycanthus floridus*

We wholeheartedly agree with Bernard M'Mahon's praise in *The American Gardener's Calendar*: "Calycanthus . . . commonly called the sweet-scented shrub, is deserving of a place in every pleasure-garden, on account of the delightful odour of its flowers."[102]

> There is a shrub peculiar to this province, that bears a small flower, which the inhabitants term the bubby flower, it resembles that which grows on clover grass, and has peculiar qualities, for it retains its grateful and odiferous perfume for a length of time after being gathered, and as it withers, encreases; the name given to the flower arises from a custom that the women have of putting this flower down their bosoms, letting it remain there till it has lost all its grateful perfume.
>
> —Anburey, *Travels*, 2: 396

> Calycanthus, one of the most charming and highly prized shrubs of the southern United States. . . . The wood and the leaves [and of course the blossoms] have an aromatic fragrance, and since it is a shrub of beautiful foliage, it is used in South Carolina [in extensive gardens] to form hedges and the little alleys of gardens.
>
> —Castiglioni, *Viaggio*, 365–66

FIG. 49. Sweet Scented Shrub

Attributed to John Laurens; Special Collections, University of South Carolina Libraries

The Basic Orchard

The Records of the Moravians in North Carolina listed fruit trees "that have been brought here, and that do well," including "very many" apple and peach trees; "a few" pear, apricot, sweet cherry, sour cherry; and also quince, white grapes, gooseberries, currants. Traveling widely in the mid-Atlantic region, Peter Kalm reported "orchards have seldom other fruit than apples and peaches. Pear trees are scarce. . . . They sometimes have cherry trees" while in some areas "every countryman, even the poorest peasant, had an orchard with apples, peaches, chestnuts, walnuts, cherries, quinces and such fruits." Fifty years later, in 1802, naturalist François Michaux noted that although settlers on the southern frontier found the climate generally favorable to fruit trees, "They have confined themselves, at present, to the planting of peach and apple trees."[103]

> [William Bartram] observed in the ancient cultivated fields . . . [persimmon, honey locust, plum, beauty berry, red mulberry, walnut, hickory] natives of the forest, yet they thrive better, and are more fruitful, in cultivated plantations, and the fruit is in great estimation with present generation of Indians.
>
> —Bartram, *Travels*, 57

> Fruit is certain in the coldest part of North-Carolina, every third or fourth year. This is sufficient encouragement for the culture of an orchard, for the farmer may then . . . distill all that will be necessary for family consumption until the next fruit year.
>
> —*North-Carolina Chronicle; or Fayetteville Gazette*, September 27, 1790

Apple. *Malus pumila*

William Byrd, surveying the Virginia–North Carolina boundary line in 1728, observed that "those who take care to plant good Orchards are . . . Industrious People." On the other hand, "in that Indolent climate; nor is it at all Strange that such improvident People, who take no thought for the Morrow, shou'd save themselves the Trouble to make Improvements":

> Tho' if [those negligent people] could trust futurity for any thing, they certainly would for Cyder, which they are so fond of, that they generally drink it before it is done working, lest the Fermentation might unluckily turn it Sowr.
>
> —Byrd, *Histories*, 110

Apple and pear grafts were listed for sale in a 1746 *South Carolina Gazette*. In March 1754 North Carolina Moravians "laid off the orchard [and] . . . planted apple and peach trees." Around the house at a Quaker plantation in Tennessee in 1802, "magnificent apple-trees were planted, which, although produced from pips, bore fruit of an extraordinary size and luxuriance in taste, which proves how well this country is adapted for the culture of fruit trees."[104]

> Aug 4, 1775, Salem, N.C. Today two wagon-loads of apples were sold here, they came from the South Fork and the Yadkin. Yesterday a wagon-load from Bethabara was sold here.
>
> —*Records of the Moravians*, 2: 880

About two hundred miles from the sea-coast the apple trees are magnificent, and in the county of Lincoln several Germans make cyder.

—Michaux, "Travels 1801–1802," 298–99

In the time of pressing cyder, if a traveller should call . . . he is generally offered as much fine cyder as he can drink, is frequently requested to stay all night, and made heartily drunk in the bargain if he chooses it.

—Smyth, *Tour*, 70

The best sorts of Apples serve at the last course for the table, in most men's houses of account, where, if there grow any rare or excellent fruit, it is then set forth to be seene and tasted.

Divers other sorts serve to bake . . . either in pyes or pans, or else stewed in dishes with Rosewater and Sugar, and Cinamon or Ginger cast upon.

Some kinds are fittest to roast in the winter time, to warme a cup of wine, ale or beere; or to be eaten alone, for the nature of some fruit is never so good, or worth the eating, as when they are roasted. . . . Some sorts are best to make Cider of.

—Parkinson, *Paradisi*, 588–89

[Pomade, pummatum, pomatum:] There is likewise made an ointment with the pulpe of Apples and Swines grease and Rose water, which is used to beautifie the face, and to take away the roughnesse of the skin, which is called [in] shops Pomatum: of the Apples whereof it is made.

—Gerard, *Herball*, 1460

There is a fine sweet oyntment made of Apples, called Pomatum, which is much used to helpe chapt lips, or hands, or for the face.

—Parkinson, *Paradisi*, 588–89

Note: Despite the name (*pomum*, Latin for "apple"), not every pomade included apple, some being only a combination of greases, oils, and/or waxes with one or more ingredients added for scent or benefit.

Today we don't have to worry about ways to keep the (now extinct) Carolina parakeets from destroying our apples. However, the following advice may yet apply for discouraging starlings or squirrels.

The havoc they [Carolina parakeets] make is sometimes so great that whole orchards are laid waste, in spite of all the noises that can be made or mawkins [scarecrows, malkins] that can be dressed up to frighten them away.

—Byrd, *Histories*, 206

Peach. *Prunus persica*

Traveling through the Carolinas in 1709, John Lawson noted "a great many sorts of this Fruit, which all thrive to Admiration. Peach-Trees coming to Perfection (with us) as easily as the Weeds." A hundred years later, François Michaux found "here [in South Carolina], as well as in Tennessea, and the greatest part of Kentucky, they cultivate no other but the peach . . . on account of their making brandy with it."[105]

There are five or six species of them, some forward, and others late, of an oval form, and much larger than our garden peaches. All the peaches grow in the open field, and proceed from kernels without being either pruned or grafted. They shoot so vigorously, that at the age of four years they begin to bear. The major part of the inhabitants plant them round their houses, and others have great orchards of them planted crosswise. They turn the hogs there for two months before the fruit gets ripe. These animals search with avidity for the peaches that fall in great numbers, and crack the stones of them for the kernels.

—Michaux, "Travels 1801–1802," 242

[Near Richmond, Virginia, July 1779] the peach trees present their fruit, to allay the parching thirst the heat occasions; and it is deemed no trespass to stop and refresh yourself and your horse with them; if the owner of the plantation perceives you gathering, he will come and direct you to the tree that bears the best fruit, for the peaches of this country are of various sorts, and though they are in such abundance, that they feed the hogs with them, yet there are very few except in gentleman's gardens, which have that flavor and juice of those that grow in England.

—Anburey, *Travels*, 2: 237

When apricot, peach, and nectarine-trees, are in bloom, some of the choice kinds should be defended from frost, if it should happen at that time, by covering the trees with mats. . . . To preserve the blossoms of young fruit, I have sometimes stuck the trees with branches of ever-greens, which I have found to be often serviceable.

—Mawe, *Every Man His Own Gardener*, 110–11

An infusion in water . . . of the fresh-gathered [or dried peach] flowers . . . sweetened with sugar, proves for children an useful laxative.

—*Edinburgh Dispensatory*, 250

Peach blossom syrup and decoction of peach leaves appear in numerous receipts for treating everything from quinsey to consumption to dry gripes in Tennent's *Every Man His Own Doctor*. Although we cannot attest to the success of these interesting home remedies, certainly peach blossom syrup seems a pleasant vehicle for administering medicine.

For Collick

Drink a Quarter of a pint of the Decoction of Peach Leaves, with 2 Spoonfulls of Syrup of Peach Blossoms: And this Purge you ought to repeat 2 or 3 times to carry off all the remains.

—Tennent, *Every Man His Own Doctor*, 23

The peach blossoms were also utilized for their delicate flavor. Those who wish to collect peach blossoms today should take care not to collect in areas where pesticides may have been applied. In the nineteenth century we find South Carolina cooks suggesting peach water in place of rose or orange flower water. The following receipt for peach blossom infusion and syrup had been published in England a century earlier.[106]

To make Syrup of Peach Blossoms

Infuse Peach Blossoms in hot Water, as much as will handsomely cover them. Let them stand in Balneo [bain-marie, a hot-water bath], or Sand, for twenty-four Hours covered close; then strain out the Flowers from the Liquor, and put in fresh Flowers. Let them stand to infuse as before then strain them out, and to the Liquor put fresh Peach-Blossoms the third time, and if you please, a fourth time. Then to every Pound of your Infusion, add two Pounds of double-refined Sugar, and setting it in Sand or Balneo, make a Syrup, which keep for Use.

—Glasse, *Art of Cookery* (1747), 153–54

Peach-leaf Brandy

Put some green peach leaves in a jar, pour on as much white brandy as will cover them, close the jar and let them steep for three days; then take them out and fill the jar with fresh leaves. Repeat this process till you have a very strong infusion and then strain it and put the liquid up in bottles. A little of this brandy is excellent for flavoring cakes, puddings, &c.

—Bryan, *Kentucky Housewife*, 409

The Carolina Housewife's "Jumbles" and Mrs. Ashburn's "Little Cakes" are good basic receipts that use flavorings featuring rose or peach or other herbs. Since these early receipts do not contain the chemical leavening agents that contribute a characteristic bitterness to today's baked goods, the delicate flavors of herbal infusions may be appreciated.

Jumbles. No. 3

One pound of butter, two pounds of white sugar, one stick of cinnamon, one nutmeg, one glass [one-fourth cup] of rose (or peach) water, the yolks of eight eggs; rub the butter and sugar to a cream; beat the eggs very light, and add them to the butter and sugar; then the spice and essence; then stir in enough well dried wheat flour to make the mixture stiff enough to roll; twist your jumbles, dip them in powered sugar [extra fine sugar, not confectioners' sugar], and bake on a tin sheet.

—Rutledge, *Carolina Housewife*, 202–3

Jumbles (Interpretation)

Halving the above receipt produces a more reasonable quantity of dough.

Cream together:

one-half pound butter and two cups sugar

Mix in:

four egg yolks

one-half teaspoon each nutmeg and cinnamon

two tablespoons rose (or peach) water, or flavored brandy

Add:

flour enough to form a manageable dough (up to two and a half cups)

Roll out and cut or twist into shapes and bake in medium oven as other cookies.

Peach kernels were also employed as a flavoring for cakes and puddings. The somewhat toxic kernels inside peaches and other stone fruits were prized for their strong almond flavor. English cookbook author Richard Bradley wrote of the cordial ratafia, "If you cannot get Apricot or Peach-Stones . . . you may use the Kernels of Plum-Stones, Cherry Stones, or Prunes, and they will make little difference. . . . Break the Shells, and put in Shells and all."[107]

Caution: Stone fruit kernels contain cyanide and should be avoided. Nevertheless, ratafia appears to have been a widely enjoyed cordial for sipping as well as flavoring. Ratafia recipes have been passed down from a range of times and places, as illustrated below, one method published in early-eighteenth-century England, the second from a backcountry North Carolina cookery manuscript a century later.

To make Ratafia

Take the Kernels of Apricots [or Peaches] . . . bruise them a little; then put them into three or four Quarts of Brandy, and let them steep four or five Days: then strain them off, and add as much fine Sugar powder'd [extra-fine sugar, not confectioners' sugar], as will make it sweet to your Taste.

—Bradley, *Country Housewife*, 2: 70–71

A Fine Cordial

Add to 1 Gallon of old peach brandy 4 oz almonds beaten in a marble mortar, do. [ditto] peach kernels blanch'd & beaten 2 lb Sugar, ½ Gallon water shake these often during 3 weeks then filter thro' paper, & bottle it.

—Cameron, "A Friend to the Memory, Recipes"

This flavoring source, of questionable safety, was employed by early cooks, as seen in this South Carolina receipt:

Little Cakes Mrs Ashburn

Take half a pound of flower half a pound of Sugar beat fine a quarter of a pound of Butter a Teaspoonful of Ratifie & a few Currants mix them well together roll them in round Balls put them on tins to bake them directly.

—Unknown author, Herbal and household book

And a Great Variety of Wild Fruits, Nuts, and Berries

Seasonally abundant wild fruits and berries provided welcome variety at frontier tables. The rich berry harvest began with strawberries in late spring and progressed through the summer with blueberries, raspberries, blackberries, mulberries, and others. The luscious flavors were enjoyed throughout the year as preserves, pickles, and spirits. Following the berry season, nature provided a harvest of grapes as well as a variety of nuts.

Nuts.

Black Walnut. *Juglans nigra*
Hickory. *Carya* species
Chestnut. *Castanea dentata*

The American black walnut was not enjoyed for food as the English or French walnut was, but it was prized as "a very good and durable Wood." John Lawson also reported that the walnuts contained a useful oil. "The Nuts have a large Kernel, which is very oily, except lain by, a long time, to mellow." The walnut hulls were widely employed as a dark brown dye.[108]

Black Walnut is the best and most valuable wood in this country for joiner's work; it is a beautiful brown. The nuts are indeed walnuts, with outer hulls which are used in dyeing, but the nuts cannot be opened with a knife.

—*Records of the Moravians*, 2: 560

They have a sort of walnut they call black walnuts, but are very rank and oily, having a thick, hard, foul shell, and come not clear of the husk as the walnut in France doth; but the inside of the nut, and leaves, and growing of the tree, declare it to be of the walnut kind.

—Beverly, *History of Virginia*, 2: 14

With the bark of the Black walnut and the husk of its fruit are dyed wool and dark-colored textiles.

—Castiglioni, *Viaggio*, 393

Wildlife, humans, and livestock competed for these resources. Peter Kalm reported Indians and European Americans "take great pains to find out . . . store-holes [of the squirrels], whether in trees or in the ground, as all the nuts they contain are choice."[109]

FIG. 50. Chestnut

Elizabeth Blackwell, *A Curious Herbal*, 1739

Among other plentiful nuts, "Hiccory Nuts have very hard shells, but excellent sweet Kernels," and "the old Hogs that can crack them, fatten themselves, and make excellent pork." Acorns and chestnuts provided important additional sustenance.[110]

> The Walnuts, the Hickory-Nuts . . . contains [*sic*] a vast deal of Oyl, that might be press'd out in great abundance. . . . The Trees grow very kindly, and may be easily propagated, they bear plenty of Nuts every year, that are now of no other use in the World but to feed Hogs.
>
> —Byrd, *Histories*, 208

> [Shellbark hickory:] I have seen above an hundred bushels of these nuts belonging to one [Creek Indian] family. They pound them . . . cast into boiling, water, which [after straining] preserves the most oily part of the liquid: this they call by a name which signifies hiccory milk; it is as sweet and rich as fresh cream, and is an ingredient in most of their cookery.
>
> —Bartram, *Travels*, 57

> Hickory nuts are of several sorts. . . . Some of these nuts are inclosed in so hard a shell, that a light hammer will hardly crack them; and when they are cracked, their kernel is fastened with so firm a web, that there is no coming at it. Several other sorts I have seen with thinner shells, whose kernels may be got with less trouble . . . they are all of the walnut kind.
>
> —Beverly, *History of Virginia*, 2: 14

The native American chestnut, which grew "up towards the hilly parts," was considered more desirable than the hard-shelled black walnut and hickory. Lawson described the chestnut as "smaller than those of Portugal, but sweeter"; and the chinquapin even sweeter than the chestnut though smaller.[111]

> [The chestnut tree] may be propagated by planting the nuts, which should be left in the burr, as a defence against field mice, the Spring is recommended as the best time for planting.
>
> —*Valuable Secrets*, 227

> These are not used medically tho' much in food. . . . Occasions a Load at the Stomach . . . if too many be eat of them at a time. . . . [Gastrointestinal distress is] prevented by roasting them.
>
> —Quincy, *English Dispensatory*, 117

> The nut is good raw, and is generally eaten that way. One expects to find the tree on the poorest land, the wood is good for shingles, rails and fences, as it weathers well; it is not good firewood on account of the sparks it throws out. A palatable coffee can be made from the chestnuts.
>
> —*Records of the Moravians*, 2: 560

The American chestnut, once a prominent tree in our forests, was largely erased from this landscape by a blight in the early twentieth century. Some chestnut trees continue to sprout and grow for several years and even bear nuts before again falling prey to the blight.

Grapes. *Vitis vinifera*
Muscadine. *Vitis rotundifolia*

Muscadines were prized, although other wild grapes were disappointing. Early explorers and settlers had high hopes for cultivating grapes that did not pan out in those early days.

> As for that noble Vegetable the Vine, without doubt, it may . . . be improved, and brought to the same Perfection as it is . . . in the same Latitude in Europe.
>
> —J. Lawson, *New Voyage*, 117

The Grapes we commonly met with were black, tho' there be two or three kinds of White Grapes that grow wild. The Black are very Sweet, but Small. . . . I have Drunk tolerably good Wine prest from them, tho' made without Skill. There is then good Reason to believe it might Admit of great Improvement, if rightly managed.

—Byrd, *Histories*, 196

A small kind of wild grapes . . . when they are quite ripe are eaten raw, and have a very good flavor, being a mixture of sweet and acid. Some people dry these grapes and bake them in tarts, etc. They likewise use them for dried sweetmeats [also to make wine, vinegar, and brandy].

—Kalm, *Travels*, 1: 199–200

In 1756 the Moravians in piedmont North Carolina found grapes that "grow wild in the woods, are small, but of a fairly good flavor." By 1764 they had begun "a small vineyard" of white grapes that was expanded the following year with "a bundle of grapevines" brought from eastern North Carolina.[112]

The Whole Country is Covered with pretty good wild grapes . . . it would be worth while to plant Vine Yards there. . . . Government has promised a reward of five hundred pounds to the first who shall make a Hogshead of Good Wine.

—Collet, "A Map-Maker's View of Anson County in 1769," 277

[Wild grapes] had a pleasant sourish taste. The country people gather them in great quantities and sell them in town. They are eaten without further preparation, and are commonly offered to guests.

—Kalm, *Travels*, 1: 80

[Vineyards were attempted, but] the success did not answer the expectation; only four or five various kinds survived [out of twenty-five planted in Kentucky by the beginning of the nineteenth century].

—Michaux, "Travels 1801–1802," 208

FIG. 51. Herbs "dryed, infused, or even pickled"

Teresa Armour, artist, in *The Backcountry Housewife*

A Salmagundi of Methods

A Bundle of Sweet Herbs

Early receipts typically called for a "bundle" or "faggot" of sweet herbs without specifically naming those herbs. Not all those termed "sweet herbs" are what we would consider sweet, but rather more aromatic or assertive. They do sweeten the pot with flavor, however, whether bundled or loose.

Parsley and thyme were the most commonly mentioned culinary herbs throughout the English and American sources we studied. Other favored sweet herbs included marjoram and savory to season a broth or made dish (a special dish cooked with multiple ingredients and dependent on the cook's skill). Tarragon, sorrel, spinach, mint, burnet, saffron, or marigold petals might enter the mix. Bay leaves, rosemary, and basil were less customary in eighteenth-century kitchens. We puzzled over the absence of oregano in period receipts until we realized that their winter marjoram or pot marjoram was the herb we now call oregano.

It is clear that national origins shaped the American palate. A German cook was likely to add chervil to the expected sweet herbs. French receipts recommended gentler seasonings for broths and sauces—often only parsley and chives, occasionally bay leaves or thyme, sorrel to brighten the flavor, or perhaps spinach to "green" a dish. The Frenchman Louis Ude reached culinary heights with "a large bunch of all such sweet herbs as are used in the cooking of a turtle: sweet basil, sweet marjoram, lemon thyme, winter savory, two or three bay leaves, common thyme, a handful of parsley and green onions, and a large onion stuck with six cloves," plus cayenne to be added later in the cooking.

A general note on eighteenth-century seasonings: The sweet herbs did not stand alone but were generally boosted with spices. Pepper, mace, and nutmeg were favored, along with cloves, cinnamon, ginger, or allspice. Some receipts called for the sharpness of mustard, radish, horseradish, or nasturtium. One of the sorrels or verjuice might stand in for vinegar or lemon.

Although fresh herbs were generally preferred, various methods were employed to capture their essence. Herbs might be dryed, infused, or even pickled for out-of-season use.

Dried Herbs

Scotswoman Margaret Dods directed: "Parsley may be picked, and dried by tying it in bundles to a rope, or drying it in a cool oven; and so may other herbs. . . . The herbs which are generally kept dry, are mint, marjoram, thyme, sweet basil, and sage." Certain blossoms, as well as leaves, may be successfully dried.[1]

Having well dried them [leaves and flowers], put them up in Brown Paper, sewing the Paper up like a Sack, and press them not too hard together, and keep them in a dry Place near the Fire.

—Culpeper, *English Physician*, 368

Gather chamomile, marigold, and such other flowers as may be wanted, and that are now in bloom; spread the flowers in the shade till sufficiently dry, and then put them up in paper bags.

—M'Mahon, *American Gardener's Calendar*, 429

Powder of Fine Herbs

for flavouring soups and sauces, when the fresh herbs cannot be obtained

Dry, in summer, parsley two ounces[;] of lemon-thyme, winter savory, sweet marjoram, and basil, each an ounce; lemon-peel dried, an ounce. Dry slowly and thoroughly; pound, then bottle.

—Dods, *Cook and Housewife's Manual*, 254

To keep the flavor of sweet-marjoram; the best of all herbs for broth and stuffing. It should be gathered in bud or blossom and dried in a tin oven at a moderate distance from the fire; when dry, it should be immediately rubbed, sifted, and corked up in a bottle carefully.

—Child, *American Frugal Housewife*, 37

Salads and Their Dressing

The eighteenth-century salad could be as simple as a single vegetable or a surprising and delightful combination of vegetables, raw or boiled or pickled; fruits; boiled eggs; and/or meats. Season and availability dictated ingredients. The salad did not necessarily include leafy vegetables. The ultimate salad, the salmagundi, included a little of everything favored, in an artistic arrangement.

People have their own favoured methods of making all kinds of salads from herbs and roots, cooked, raw or pickled, combined with meat, fish or salted products.

—Von Rumohr, *Essence of Cookery*, 164

Basic salad dressings were oil and vinegar (or butter, with or without vinegar, for warm salads) with salt and pepper. Chopped boiled egg might be added for body. Those who could buy it used imported olive oil; bear oil (processed from bear fat) was considered a good alternative. John Evelyn's early English directions for dressing a salad do not seem at all foreign today:

Dressing & Sauce [for Salad]

The old Recipe is that

A Wise man should gather the Herbs
An Avaricious [greedy] man fling in y[e] salt & vinegar:
A Prodigal [extravagant person] the Oyle.

From Aprill to September, leave out Horse-Radish & Ginny [Guinea] pepper [related to African black pepper, *Piper* species]. From October to March Take of cleare unsmelling salade oyle 3 parts, of Vinegar or Orange juice one part, letting some horse radish steepe in it, straine it in another part of vinegar & bruse a pod of Ginny pepper in it, straine it off also, and add as much grated Tewksberry mustard as will cover halfe a crown; then add the juice of a lemmon, a little salt, beate all well together: But put not in the Oyle & Vinegar 'til the salad is to be eaten, & then beate them all together with 2 yolkes of an Egg, boyled somewhat hard . . . the acid juices of the limmon & vinegar . . . must be more or lesse sharp, as the Company please.

—Evelyn, *Directions for the Gardiner*, 119

The English took advantage of two plants growing wild in the countryside. Grated horseradish was added to ground mustard seed to produce a particularly sharp mustard named for the town of Tewkesbury.

Herb Sauces

Although sauces and soups are best made with fresh herbs, when those are unavailable dried herbs can also elevate flavors, particularly in winter when basic ingredients are limited and perhaps not at their prime. The following receipts offer the cook unlimited herb combinations for tasty sauces.

Fines Herbs

Take a cup full of chopped shalots or onions, a little parsley, one laurel-leaf [bay leaf], estragon [tarragon], thyme, sweet marjoram and lemon-peel; chop them fine, and put them, with three ounces of butter, in a stew-pan over the fire; stir the mixture a short time, until the butter becomes clear; then put it in an earthen pot, where it will keep eight days sound.

—Rutledge, *Carolina Housewife*, 219

A great variety of delicious sauces can be made, by adding different herbs to melted butter. . . .

To melt butter . . . weigh one quarter of a pound of good butter, rub into it two teaspoonsful of flour; when well mixed, put it in the sauce pan with [two] table spoonsful of water, and a little salt; cover it, and set the sauce pan in a larger one of boiling water. . . .

To begin with parsley—wash a large bunch very clean, pick the leaves from the stems carefully, boil them ten minutes in salt and water, drain them perfectly dry, mince them exceedingly fine, and stir them in the butter when it begins to melt. . . . Chervil, young fennel, burnet, tarragon, and cress, or pepper grass, may all be used.

—Randolph, *Virginia Housewife*, 112–13

[A Cold Herb Sauce]

To make a cold herb sauce, take one half part spinach and one quarter tarragon, the last quarter being a mixture of sorrel, purslane, parsley and chervil. Add a tiny shallot, or half of a larger one, with a few basil, marjoram and thyme leaves [also salt]. Pound the mixture in a mortar. . . .
Put them with boiling vinegar through a fine hair sieve into a clean earthenware pot. . . . Bring it just to a boil. . . . The sauce may be thickened with olive oil, egg yolks, or

FIG. 52. Saucepan "some fine herbs in it"

Teresa Armour, artist, in *The Backcountry Housewife*

a little semi-solidified white stock [chicken stock], as appropriate.

—Von Rumohr, *Essence of Cookery*, 159–60

[Common Tavern Sauce]

A common browning made with butter and flour, moistened with a little broth or gravy, and some fine herbs in it, will answer the purpose of those who know no better.

—Ude, *French Cook*, 12

Soups

Springtime herbs had a place in soups as well as salads and greens. Von Rumohr noted that "in spring time it is worthwhile making plenty of soups from chopped herbs, varying the mixtures to obtain more bitter, or more strongly flavoured results." These might be simmered in water or meat broth and thickened with egg yolk or perhaps oatmeal as in the "Pottage of Chopped Herbs" below.[2]

To make Pottage of Chopped Herbs

Mince very fine, Spinage, Chives, Parsley, Marigold-flowers, Succory [Chickory], Strawberry and Violet Leaves, stamp them with Oat-meal in a Bowl, put chopped Greens with it: you may either put Broth or Water in them, if Water, boil a good Piece of Butter in it; put Sipets [pieces of bread] in the Dish, and pour it over them.

—Cleland, *New and Easy Method of Cookery*, 14

A Foundation for Herb Soups

Take a quantity of good Herbs, such as Cherville, Spinage, Sallery, Leeks, Beet Cards [stalks], and such like, with two or three large Crusts of Bread, some Butter, a bunch of sweet Herbs, and a little Salt; put these, with a moderate quantity of Water, into a Kettle, and boil them an hour and half, and strain off the Liquor thro' a Sieve, and it will be a good Foundation for Soups, either of Asparagus Buds, Lettuce, or any other kind, fit for Lent or Fastdays. These Herb Soups are sometimes strengthened with two or three Yolks of Eggs, a little before they are serv'd to the Table.

—Bradley, *Country Housewife*, 1: 26

Potage à la Julienne

Cut up fine some carrots and turnips, some onions cut in slices, also some leeks and a few stalks of celery. Chop up coarsely some lettuce, a handful of sorrel and half the quantity of chervil. Heat the roots on a frying pan with butter—not fry them exactly. When they are heated

FIG. 53. Soup Pot "add a bundle of sweet herbs"

Teresa Armour, artist, in *The Backcountry Housewife*

through, moisten the whole with good broth, and put it on the fire for at least an hour. When the julienne is done, serve it up with sippets of bread.

—Rutledge, *Carolina Housewife*, 48–49

These springtime and Lenten soups sound delicious. Soup ingredients were often less interesting in winter, however. The winter soup pot would certainly profit had the householder dried a supply of favored herbs.

Onions, carrots, parsnips, and turnips were basic seasoning agents often accompanying common sweet herbs. These roots were generally cultivated in quantities requiring a larger space than the kitchen garden allowed. Either kitchen garden or truck patch might accommodate root vegetables, legumes, and the more robust herbs (celery, artichokes, greens, squash, and cucumbers).

FIG. 54. Tea "best way to use Flowers and Herbs"

Teresa Armour, artist, in *The Backcountry Housewife*

Beverages

Herbs were typically incorporated into tea, broth, or spirits, perhaps into milk for a posset as suggested under **Woodsorrel**, **Marigolds**, and **Feverfew**. The Moravians in North Carolina found that "the best way to use Flowers and Herbs is to infuse them in Rum, Wine, Beer, or in hot Water like Tea, and drink of them two or three Times a Day."[3]

We believe the best way to enjoy any edible herb is to nibble it freshly picked in the garden; perhaps the next most obvious and simple way to enjoy an herb, fresh or dried, is infused in water to brew a tea.

I observe, that in making such Infusions as Teas of dried Herbs, the best way is to pour boiling Water upon them, and in half a Minute, at most, pour out the Water again from the Herbs . . . such Tea will then be of a fine green Colour, and full of Spirit: but if the Herbs stand longer with Water upon them, the Water will change of a brownish Colour, will lose the fine Flavour of the Herb, and become ill-tasted.

—Bradley, *Country Housewife*, 1: 92

A posset is a warm beverage of spiced milk or cream thickened with bread, grain, or eggs (three yolks per quart of liquid); to that, wine, ale, or brandy is added to form a soft curd. While warm milk and wine may sound less than delicious, you are likely to be surprised, as we were, on testing posset receipts. Madeira or sherry seemed an appropriate wine choice and was favored by several of our experimenters; others preferred a dry white wine that allowed spice or herb flavors to come forward. A loosely packed cup of woodsorrel (leaves, stems, flowers, and seeds) deliciously infuses two cups of milk for posset. (Proportion for spice or herb will vary according to strength of their flavor.) Each member of our experimental posset party enjoyed discovering this comforting warm ancestral beverage and adapting Elizabeth Raffald's eighteenth-century receipt:

To make a Wine Posset

Take a quart of new milk [perhaps herb-infused] and the crumb of a penny loaf,[4] and boil them till they are soft. When you take it off the fire, grate in half a nutmeg and sugar to your taste. Then put it into a china bowl and put in a pint of Lisbon wine carefully a little at a time, or it will make the curd hard and tough. Serve it up with toast and butter upon a plate.

—Raffald, *English Housekeeper*, 159

A Basic Wine Posset (Interpretation from Raffald)

Heat 2 cups milk with:

¼–½ teaspoon nutmeg; or 1 cup fresh woodsorrel (oxalis); or other flavoring

Stir in:

½ cup sugar and 1 cup heated wine

Add:

4 ounces bread crumbs, or 3 egg yolks

Stir constantly, just until thickened (do not allow to boil).

Flavored Wines, Spirits, and Cordials

English cookbook authors Glasse and Raffald favored "waters" flavored not only with rose petals but also elderflowers or lavender blossoms; "Mint, Baum, Parsley, and Pennyroyal Water, distill the same way." Mid-eighteenth-century southern plantation mistress Eliza Pinckney penned a letter mentioning that her daughter Harriott "had people out gathering simples . . . and is distilling herbs and flowers." *The Carolina Housewife*, published years later, suggested peach water for flavoring cakes and puddings.[5]

To Distil Red Rose-Buds [to make rose water and other herb waters]

Wet your Roses in fair Water; four Gallons of Roses will take near two Gallons of Water, then still them in a cold Still, and take the same stilled Water, and put it into as many fresh Roses as it will wet, then still them again.

—Glasse, *Art of Cookery*, 159

"Cordial waters" were distilled by an apothecary or in the kitchen of a well-equipped household. Early-nineteenth-century Scotswoman Margaret Dods noted: "As this [distillation] is a troublesome process, they

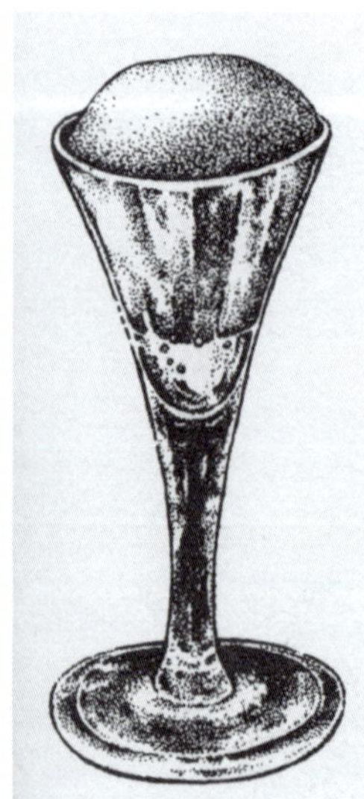

FIG. 55. "A Cordial . . . easily produced by infusing"

Teresa Armour, artist, in *The Backcountry Housewife*

[liqueurs or cordials] are generally made in small families, by infusion, which succeeds very well."[6]

A cordial is most easily produced by infusing the desired herb in brandy or whisky rather than by distilling. Cordials were administered as medicines, employed in the kitchen as flavoring agents, or sipped as pleasant, healthful beverages. The herb, plus a little sugar, was simply immersed in brandy or whisky and allowed to infuse for a week or two before straining; or, even more conveniently, a previously prepared syrup of the herb was added to spirits. Mint, balm, saffron, peach kernels, savory, rosemary leaves or flowers; also flowers of roses, violets, borage, bugloss, marigolds, orange, and peach commonly entered into herbal spirits.

> Spirit that the Herbs are put into, must be cold, or used without any Fire at all, and the Herbs in this case may be used either green or dry; here they may stand several days before the spirit that they are infus'd in be drawn off.
>
> —Bradley, *Country Housewife*, 1: 93

> The principal design of these cordials is to preserve the juice, to use when the fresh materials cannot be procured, as it is much cheaper than to use the essence, and the flavor is equally as nice.
>
> —Bryan, *Kentucky Housewife*, 392

> ***To make a Cordial Water***
>
> Take 1 qtt of Spring watter a handfull of Bame & mint & marygold flowers & leaves & put them in a Pipkin & infuse them over a Soft fier. . . . A lettle before you take it of the fier add to it ¼ ptt Juce of Oranges then Strain it Then add to 16 lb [cost or weight?] Sugar: make it into a thin Syrup then let it stand till it be Cold add to it 1 lb worth [currency] saffron tied up in a thin rag: put this into two Quarts Brandy infuse in the Brandy ½ quart peel of 1 orange.[7]
>
> —Unidentified cookbook, circa 1700, in Harbury, *Virginia's Cooking Dynasty*, 222

Herb-infused wines or spirits served several purposes. Flavored beverages can offer pleasurable and interesting sipping, of course, but these alcohol infusions are handy flavoring agents for all sorts of foods, from savory to sweet.

> ***[Vinegar or Brandy Infused with Thyme, Mint, or Other Herbs]***
>
> A very delicious flavour of thyme may be obtained by gathering it when in full perfection; it must be picked from the stalks; a large handful of it put into a jar, and a quart of vinegar or brandy poured on it, cover it very close, next day take all the thyme out, put in as much more, do this a third time, then strain it; bottle and seal it securely. This is greatly preferable to the dried thyme commonly used, during the season when it cannot be obtained in the fresh state. Mint may be prepared in the same way. The flavour of both these herbs must be preserved by care in the preparation; if permitted to stand more than twenty hours in the liquor they are infused in, a coarse and bitter taste will be extracted, particularly from mint.
>
> —M. Randolph, *Virginia Housewife*, 115

> Wines may be impregnated with the flavour of roots and herbs in the same manner as vinegar, and this generous fluid extracts even more of the flavour. The proportions for Eschalot wine, Tarragon wine, Basil wine, &c. are the same as when these herbs are steeped in strong vinegar.
>
> —Dods, *Cook and Housewife's Manual*, 258

Eschalot [shallot] wine
To four ounces of eschalots dried, chopped and pounded, or merely bruised, put a bottle of sherry. Infuse for a fortnight and strain off. If for beef only, horseradish sliced may be added, or rather substituted for part of the eschalots.
—Dods, *Cook and Housewife's Manual*, 258

A variety of things are used for perfuming wines; such as sweet herbs, peach-leaves, sweet bay-leaves, almonds, kernels of fruit, bergamot, cloves, ginger, &c. . . .
[To improve a] cheap wine of mixed fruit . . . Perfume with a quarter-pound of ginger, and a handful of sweet marjoram and lemon-thyme.
—Dods, *Cook and Housewife's Manual*, 467, 469

See **Rose** for rose brandy instructions and **Violet** for violet water.

Herb Vinegars

"Flavoured Vinegars . . . are a cheap and agreeable addition to sauces, hashes, and ragouts, and have the convenience of being always at hand, at seasons when herbs are either very costly or not to be procured." Vinegar infusions are a convenient mode of preserving almost any herb for out-of-season use—tarragon, rosemary, thyme, mint, nasturtium, shallot, capsicum peppers, and horseradish along with elderflower, clove-pinks, rose blossoms, lavender, and sage blossoms might flavor vinegars.[8]

Tarragon Vinegar
Pick the leaves of the tarragon from the stalks just before it blooms; spread them out on a cloth, and let them lie for two or three days to dry a little; then put them into a jar [with vinegar], close it, and set it by for one week; then strain and bottle it. If you wish it very strongly flavored with the tarragon, fill it up the second time with fresh leaves and soak them in the same manner. Any kind of nice sweet herbs may be soaked in the same manner for the purpose of seasoning.
—Bryan, *Kentucky Housewife*, 179

[In dressing a salad use] the best Wine Vinegar . . . impregnated with the Infusion of Clove-gillyflowers, Elder, Roses, Rosemary, Nasturtium, &c. inriched with the Vertues of the Plant.
—Evelyn, *Acetaria*, 172

[Elder flower vinegar, made with alegar, or malt vinegar,] makes a pretty mixture on the side table with tarragon vinegar, lemon pickle, etc.
—Raffald, *English Housekeeper*, 176

Lavender Vinegar
Gather your Lavender blossoms & put them in a bottle, fill it up with the best vinegar & let it stand a week in the sun before you use it. [Added in another hand:] Very refreshing for sick persons.
—Purviance, Household book, 116

Pickled Herbs

Vinegar and salt were essentials for preserving all sorts of foodstuffs—meats as well as many vegetables and fruits. Directions for pickling nasturtium and radish seedpods, and cucumbers, of course, are included under those plant sections in this book. Similar receipts were ubiquitous in period cookery books. Pickling flowers was also suggested. May's 1680 *Accomplisht Cook* is one of the few sources to include sugar in any sort of pickle. May seems to have been writing for more wealthy and extravagant kitchens in that earlier century. Cleland's mid-eighteenth-century receipt for elder flower buds is in the mainstream tradition, with no sugar. Both receipts seem worth a try.

To Pickle any kind of Flowers

Put them into a gally-pot [glazed ceramic jar] . . . with as much sugar as they weigh, fill them up with wine vinegar; to a pint of vinegar a pound of sugar, and a pound of flowers; so keep them for sallets or boild meats.

—May, *Accomplisht Cook*, 138

To Pickle Elder Flowers when they are green, and before they are blown [in full bloom]

Let them ly in a strong Pickle of Salt and Water two Days. Then drain them, and put them in a Pan to green, with as much Water as will cover them, and two Gills of Vinegar [one-half cup]; put them on a very slow Fire . . . when they are green, dry them with a Cloth, then pour on them as much boiling Vinegar, with Cloves, Mace, Pepper and Ginger in it as will cover them. . . . Nasturtian Buds are pickled the same Way.

—Cleland, *New and Easy Method of Cookery*, 166

Candies, Conserves, Comfits, Pastes, Syrups, and Honeys

By the end of the eighteenth century conserves and other sweet herb preparations were considered more proper for culinary than medical use. Buchan's *Domestic Medicine*, for example, stated: "Every Apothecary's shop was formerly so full of these preparations, that it might have passed for a confectioner's warehouse. They . . . may rather be classed among sweetmeats than medicines."[9]

To candy Flowers

Gather your flowers when dry, cut off the leaves as far as the colour is good; according to your quantity, take of double refined sugar, and wet it with fair water, and boil it to a candy height; then put in your flowers, of what sort you please, [violets, borage, rosemary, also young mint leaves] . . . take them out as quick as you can, with as little of the syrup as may be . . . and with a knife spread them . . . and when they are dry from syrup, have ready some double refined sugar beaten and sifted, and strew some on your flowers; then take the flowers in your hand, and rub them gently in the hollow of your hand, and that will open the leaves [petals].

—Smith, *Compleat Housewife*, 221

The receipt above is the more delicate and tedious of the five for candying flowers in E. Smith's cookery book. "Rub them gently" is misleading. We find unfolding the petals requires patience and a toothpick. Seventeenth-century author Robert May had recommended candying an even longer list of flowers "for Sallets": violets, cowslips, clove gillyflowers, roses, primroses, borage, bugloss. "For better curiosity" one might garnish a dish by adding cucumber leaves and branches to the candied flowers. Or instead of candying one might simply whiten or frost herbs and berries with egg white and sugar.[10]

For to whiten geliflowers, roses and violets [and fennel blossom or seed branches] take the white of an egg, with a small drop of flower of orange water, beat them together,

FIG. 56. "Put it into Gallipots, tie it over with Paper, and over that a Leather"

Teresa Armour, artist, in *The Backcountry Housewife*

and steep [or dip] your flowers in it, then take them out and . . . shake them; put powder sugar [not confectioners' sugar, but extra fine] over them, and dry them near the fire.

—La Varenne, *French Cook*, 233

You have likely noted multiple references to conserves among the different herbs included in our *Curious Garden*. Conserves were prepared from the flowers of some herbs, from the leaves of others.

To make Conserve of Red Roses, or any other Flowers
Take Rose-buds, or any other Flowers, and pick them; and cut off the white Part from the red, and put the red Flowers and sift them through a Sieve to take out the Seeds; then weigh them, and to every Pound of Flowers take two Pounds and a half of Loaf-sugar; beat the Flowers pretty fine in a Stone mortar, then by degrees put the Sugar to them, and beat it very well, till it is well incorporated together; then put it into Gallipots, tie it over with Paper, and over that a Leather, and it will keep seven Years.

—Glasse, *Art of Cookery*, 153; also Smith, *Compleat Housewife*, 218

Candylike cakes, tablets, lozenges, or pastes might feature roses, marigolds, mint, caraway, horehound, or aniseed in addition to the herbs listed in the following receipts. Scots cook Margaret Dods clarified the timing for combining the stamped herbs with the boiling sugar, "mixing the flavoring agent very well with the boiling sugar, and pour it out."

To make a Paste of Violets, Cowslips, Burrage, Bugloss, Rosemary flowers, &c.
Take any of these flowers, pick the best of them, and stamp them in a stone mortar, then take double refined sugar, and boil it to a candy height with as much rosewater as will melt it, stir it continually in the boiling, and being boiled thick, cast it into lumps upon a pye plate, when it is cold, box them, and keep them all the year in a stove [a dry place].

—May, *Accomplisht Cook*, 200

To make Violet Cakes
Take the finest violets you can get, pick off the leaves, beat the violets fine in a mortar with the juice of a lemon. Beat and sift twice their weight of double-refined sugar, put your sugar and violets into a silver saucepan or tankard. Set over a slow fire, keep stirring it gently till all your sugar is dissolved (if you let it boil it will discolour your violets), drop them in china plates. When you take them off, put them in a box with paper betwixt every layer.

—Raffald, *English Housekeeper*. 120

We have found that each cup of violet blossoms (packed) combined with about four tablespoons sugar and four teaspoons lemon juice produces a gummy candy that gradually hardens. The lemon juice triggers a dramatic color change from a dull blue-gray to a lively magenta or violet.

Few kitchens (then or now) boast a silver saucepan, but Raffald's receipt for "Violet Cakes" is a reminder to use a vessel that will not discolor the delicate candies. Herbal confections "are cordial and sometimes even medicinal, and may be easily and very cheaply prepared at home." Actually, such luxuries were likely uncommon except in prosperous households.[11]

Conserves, syrups, and honeys seem more practical than candies and pastes. The varying flavors of herbs listed above for candying and conserving are also appropriate for extracting and preserving in syrup or honey.

Hannah Glasse gave an early English receipt for a syrup of peach blossoms (readily available in the South) in addition to the expected roses and clove gillyflowers. A pre-1744 Virginia cookbook manuscript and other American sources included directions for "Syrrop of Gillyflowers" and of violets, although Eliza Smith's receipt, below, is more precise.[12]

Herbal syrups or honeys are similarly prepared. Ingredients are combined in a jar, covered, set into simmering water for several hours, then strained. This process may be repeated for stronger flavor. The proportions are rather indefinite, but a range of measurements produces a sweet-scented sauce.

> *To make Syrup of any Flowers*
> Clip your flowers, and take their weight in sugar; then take a gallipot, and put a row of flowers and a strowing of sugar, till the pot is full; then put in two or three spoonfuls of the same syrup or still'd water; tie a cloth on the top of the pot, and put a tile on that, and set your gallipot in a kettle of water over a gentle fire, and let it infuse till the strength is out of the flowers, which will be in four or five hours; then strain it thro' a flannel, and when it is cold bottle it up.
>
> —Smith, *Compleat Housewife*, 215

You may easily experiment with these basic receipts for syrup and honey, testing your favorite fragrant flowers or leaves. Our favorite, rose honey, is particularly lovely.

> *To Make Honey of Roses*
> Take a pinte of honey, boyle it & scum it. Then add as many bruised leavs [petals] of red roses buds (the whites beeing cut of) as you may easily stirr it. Then cover it close & boyle the pot in water till you think the goodness of the roses is in the honey, then change the roses once or twice in the same manner, & at the last, strayn out the roses clear.
>
> —*Martha Washington's Booke*, 274

We have simplified the above receipt with good results. This method works well with smaller proportions as available scented roses and honey dictate. Even one-fourth cup of honey is enough for a good test batch with a handful of rose petals or other herbs. When using clear honey, you may choose to skip the first step of skimming the foam from the heated honey. Cutting off the white base of each rose petal seems unnecessarily tedious. With well-scented roses, the process need not be repeated. You will imagine all sorts of uses for rose honey; we delight in drizzling a bit over fresh May strawberries.

Cosmetics, Perfumes, and Air Fresheners (Not for Internal Consumption)

The dooryard garden furnished materials for a multitude of uses, from personal care preparations to air-purifying vinegars thought to clear the air of contagions. The following receipts and those we encountered throughout our research illustrate the variety to be discovered in early sources.

> *Aromatic Vinegar*
> Take of the dried tops of rosemary and dried sage leaves each four ounces, dried flowers of lavender two ounces, and two drachms of cloves; put them in a quart of distilled vinegar, macerate for one week afterwards, express the liquor and filter it.
>
> Another Mode is to steep rose leaves, pinks and lavender blossoms in equal proportions, in distilled or white wine vinegar, and express and filter as before directed.
>
> —Bryan, *Kentucky Housewife*, 436

The Kentucky Housewife added thyme to Mary Randolph's "Vinegar of the Four Thieves," with directions to sprinkle it "in sultry rooms to prevent faintness."[13]

> *Vinegar of the Four Thieves*
> Take lavender, rosemary, sage, wormwood, rue, and mint, of each a large handful; put them in a pot of earthen ware, pour on them four quarts of very strong vinegar, cover the pot closely, and put a board on the top; keep it in the hottest sun two weeks, then strain and bottle it, putting in each bottle a clove of garlic. When it has settled in the bottle and become clear, pour it off gently; do this until you get it all free from sediment. The proper time to make it is when the herbs are in full vigour, in June. This vinegar is very refreshing in crowded rooms, in the apartments of the sick and is peculiarly grateful when sprinkled about the house in damp weather.
>
> —M. Randolph, *Virginia Housewife*, 220–21

> *Oil of Flowers [for the hair]*
> Take any kind of fragrant flowers, from which you wish to extract the flavor; macerate them in the best Florence oil [olive oil], repeating the process with fresh leaves till the oil is as odoriferous as you wish it; then express the oil, filter and cork it up in phials. It is used for the hair instead of pomatum.
>
> —Bryan, *Kentucky Housewife*, 436

> Fomentations. These are generally made of the leaves or flowers of plants, as chamomile-flowers, mallows, elder-flowers, poppy-heads, wormwood, &c. They are best made by infusion in boiling water kept hot near the fire, but the vegetables may be boiled. Fomentations are generally applied by dipping flannel . . . into the boiling decoction and wringing it quickly out.
>
> —Dods, *Cook and Housewife's Manual*, 483

Miscellaneous Gardening Theories and Practices

Although this *Curious Garden of Herbs* is not meant to serve as a gardening guide, we cannot resist passing along a sampling of gardening details that may be of use to readers, whether history buffs, those actively interpreting gardens at historic sites, or folks just interested in a good tale. Many quaint and intriguing practices are waiting to be discovered in the old herbals and gardening guides.

Clearing the Land; Managing the Garden

"Taming the wilderness" is an accurate description of the task facing early settlers. The following "Thoughts on the Practice of Agriculture: for the Farmers of North-Carolina" helps us imagine those pioneer gardeners cultivating their virgin tracts.

> Trees should be girdled two or three years before opening the soil, the bark and small limbs fall and form a rich manure, and the timber can be carried off for fuel and other uses, to prevent falling on the grain, or forming a retreat for vermin after the soil is opened. The best time for opening soil is in autumn, the sods may be rotted, and the natural manures imbibed through the course of the winter. Indeed all arable ground whether new or old should be ploughed at this time.
>
> —The *North-Carolina Chronicle; or Fayetteville Gazette*, September 27, 1790

> Beds of *Sweete-herbs* should be stirrd up, and new molded [soil loosened and shaped], every second yeare in the Spring, to make them strike fresh rootes.
>
> —Evelyn, *Directions for the Gardiner*, 70

> One may sow Reddish, & Carrots together on the same bed: so as the first may be drawn, whilst the other is ready: or sow Lettuce, purselan, parsneps, carrots, Reddis on the same beds, & gather each kind in their season, leaving the parsneps to Winter.
>
> —Evelyn, *Directions for the Gardiner*, 70

> [September:] Gather all sorts of seeds as they ripen, and spread them to dry upon mats or cloths; and then beat them out of their husks, and put them up in a dry place until the seasons for sowing them.
>
> —Miller, *Kalendar*, 281

> Remember, that in weeding, "An hour in time, will save nine."
>
> —Gardiner and Hepburn, *American Gardener*, 66

Insects competed with gardeners for their cucumbers, melons, squashes, pumpkins and gourds.

> A yellow insect is often destructive to them; they must be killed by hand. . . . [T]hey may be nearly destroyed by spreading soot, wood ashes, or unslacked lime thinly over the plants in the morning before the dew goes off.
>
> —Gardiner and Hepburn, *American Gardener*, 56–58

Fences

The fence of a garden is an important matter . . . not only as giving protection against intruders, two-legged as well as four-legged, but as affording shelter in cold weather and shade in hot . . . of great utility in an American Garden, where cold and heat are experienced in an extreme degree.

—Cobbett, *American Gardener*, 13

Fences were indeed necessary around all cultivated land to keep out livestock, which roamed free. For the kitchen garden, pales (pickets), boards, or rails were closely spaced to exclude small animals and built high to discourage deer. The Moravian settlers decreed that all fences "are to be kept in good Order as possible, & if thro' any one's Neglect harm should ensue, he whose Fence is found deficient is to pay the Damage."[1]

A variety of fence forms were in use—paled, woven, and rail. Archaeological evidence from a circa 1757–82 backcountry South Carolina site near the Georgia border indicates a succession of fence types from primitive wattle to increasingly substantial post-and-pale construction.[2]

HORIZONTALLY WOVEN (WATTLE) FENCE

Next weave a stubborn woody Fence around;
And keep pernicious Cattle from the Ground;

—Virgil (from the original Latin), 306

Sharpened stakes or pales, 1 to 2 in. in diameter, were driven into the ground every 4 to 6 in. Vines and branches were then woven between them forming a lath-like fabric. . . . [In other sections] small pales or split log rails were then driven into the ground every 8 to 12 in. . . . For the remainder of the fence, pales were driven along intervals ranging between 3 and 5 ft.

—Brooks, Groover, and Smith, *Archaeology*, 181

[Near Philadelphia:] No other fences are in use but those which are like sheep hurdles.

—Kalm, *Travels*, 1: 87

A *dead hedge* made of split hickory rods [and brush] . . . with stakes of the stoutest parts of the same rods, or of oak saplings, or some such things. . . . One advantage . . . over every other sort of fence . . . is, that it effectually keeps out poultry.

—Cobbett, *American Gardener*, 22–23

VERTICALLY WOVEN FENCE

[Eastern Virginia/North Carolina:] They also set up their Pales without any Nails at all, and indeed more Securely than those that are nail'd. There are 3 Rails mortised into the Posts, the lowest of which serves as a Sill with a Groove in the Middle, big enough to receive the End of the Pales: the middle Part of the Pales rests against the Inside of the Next Rail, and the Top of it is brought forward to the outside of the uppermost. Such Wreathing of the Pales in and out makes them stand firm, and much harder to unfix than when nail'd in the Ordinary way.

—Byrd, *Histories*, 94

[Between Philadelphia and New York:] Lathes are placed vertically, but passing alternately on the outside and the inside of cross members.

—*Moreau de St. Méry's American Journey (1793–1719)*, 120

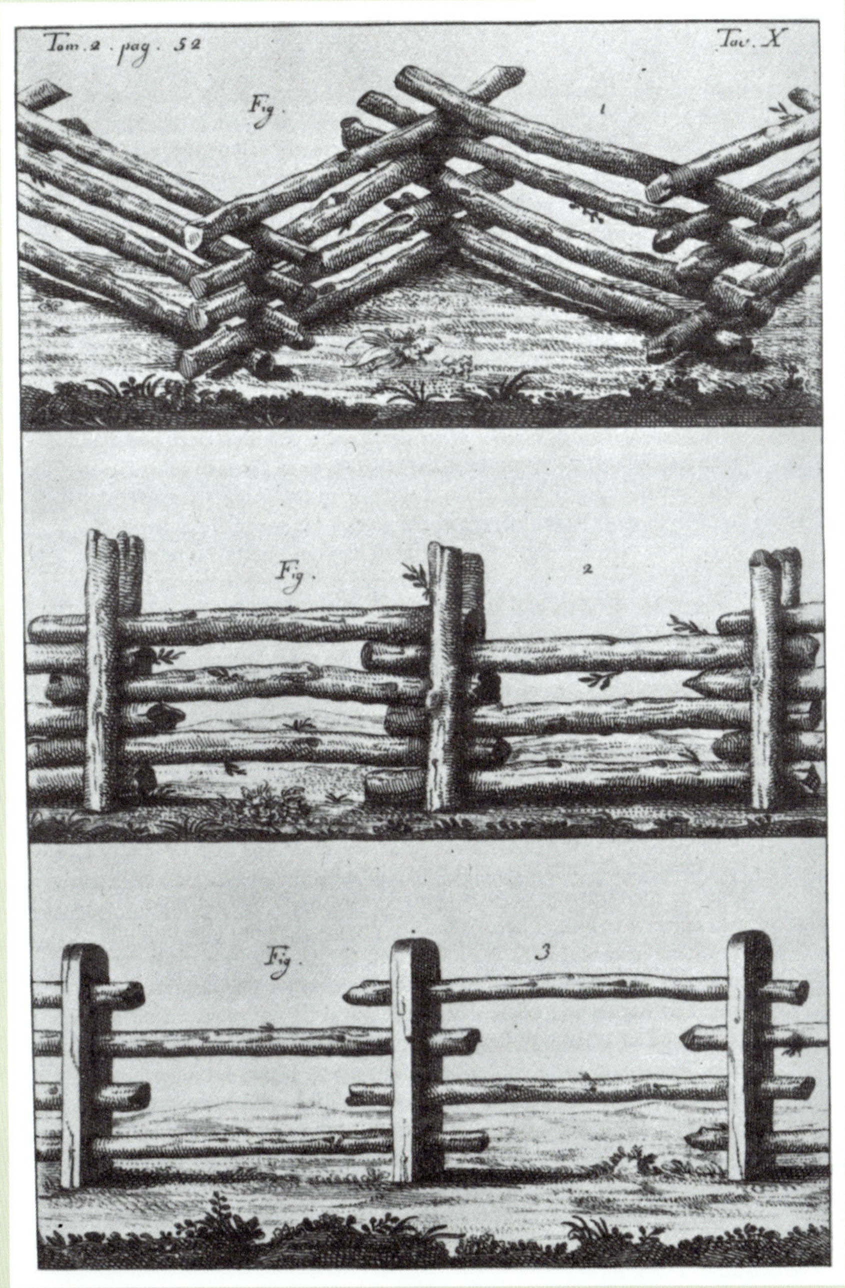

FIG. 57. Fences

Luigi Castiglioni, *Viaggio*, 1785–87. See also frontier landscapes in figures 2, 3, and 4.

POST-AND-PALE (PICKET) FENCE; BOARD FENCE

Pales for the second fence were sharpened sticks about 2 in. in diameter, while the supporting buttress posts were 6 in. or more in diameter.

—Brooks, Groover, and Smith, *Archaeology*, 181

Near towns or villages these enclosures are made with posts, fixed in the ground about twelve feet from each other, containing five mortises, at the distance of eight or nine inches, in which are fitted long spars about four or five inches in diameter. . . . The reason of their enclosing thus is principally through economy, as it takes up but very little wood.

—Michaux, "Travels 1801–1802," 134

In this country [America], board fences and palings are most in use. Rail fences ought to be avoided, as they are so easily surmounted by idle or thieving men and mischievous boys, while they admit . . . hogs and dogs at the bottom. Board fences are objectionable on account of their rapid decay.

—Gardiner and Hepburn, *American Gardener*, 104

ZIGZAG RAIL FENCE

In the Carolinas and in Virginia . . . [fences] are made in a zigzag form . . . because they require very little work. In Pennsylvania . . . and in the middle states, they are better constructed.

—Castiglioni, *Viaggio*, 319

[Southern rail fences, also called worm fences] are made of trees cut or sawed into lengths of about twelve feet, that are mauld or split into rails from four to six inches diameter. When they form an enclosure, these rails are laid so that they cross each other obliquely at each end and are laid zig zag to the amount of ten or eleven rails in height, then stakes are put against each corner, double across, with the lower ends drove a little into the ground, and above these stakes is placed a rail of double the size of the others, which is termed the rider, which, in a manner, locks up the whole, and keeps the fence firm and steady. These enclosures are generally seven or eight feet high, they are not very strong but convenient, as they can be removed to any other place . . . from a mode of constructing these enclosures in a zig zag form, the New-Englanders have a saying, "When a man is in liquor, he is making Virginia fences."

—Anburey, *Travels*, 323

[Southeastern Virginia:] The fences add to the depressing outlook. They are always made of rails split triangularly, four or five inches thick and about eight feet long. Generally the rails are placed in a zigzag manner. . . . Sometimes uprights are used to hold the crosspieces, in which case the fence is straight.

—*Moreau de St. Méry's American Journey (1793–1719)*, 69

The early wattle fence was later reinforced by a slightly more substantial fence that contained split logs placed between small posts . . . 4 in. in diameter . . . spaced at intervals ranging between 8 and 10 ft.

—Brooks, Groover, and Smith, *Archaeology*, 181

WALLS

Where they can be afforded, good brick and stone walls are not only the most ornamental and useful, but, in the end, the cheapest.

—Gardiner and Hepburn, *American Gardener*, 105

HEDGES

Between Philadelphia and New York we saw not a single growing hedge.

—*Moreau de St. Méry's American Journey (1793–1719)*, 121

[Haw-thorn] a fence, shelter and shade of everlasting duration, for a garden. . . .

And why should America not possess this most beautiful and useful plant? . . . and why not English Hedges instead of post-and-rail and board fences?

—Cobbett, *American Gardener*, 20–21

[In America, by 1826] the propagation of live hedges is every day becoming . . . of more serious importance, and will soon be one of indisputable necessity, in the long inhabited parts of the Union where timber is becoming very scarce and dear.

—Gardiner and Hepburn, *American Gardener*, 105

For gardens, hedges are advisable for two distinct purposes . . . to serve as a wall for the exclusion of trespassers . . . for the purpose of ornament and shade. . . .

For the former . . . [w]e have great varieties of native Haw-thorns . . . select whatever kind of Haw-thorn grows most luxuriously in the neighbourhood of your enclosures.

—Gardiner and Hepburn, *American Gardener*, 121

Labeling Plants

To label or not to label? In a small household garden where the gardeners knew their plants intimately, labels may not have been necessary. Even then, though, new arrivals or tender seedlings may have required identification. We found *The American Gardener's Calendar* instructions for creating labels interesting and a nice touch for the historical garden.

In severe weather, when out-work cannot be done, make labelsticks, and have them in readiness . . . made of white oak, or some other good durable wood . . . [smaller labels] may be made of old or new shingles, or pieces of good pine, cut and split . . . rub one side of the upper end for about two inches with white oil-colour paint; on which, while yet wet, write your number, or the name of the plant at full length with a black-lead pencil; this will endure any kind of weather, for one year at least.

—M'Mahon, *American Gardener's Calendar*, 51

Epilogue

The Primary Sources

The inventory of plants known to eighteenth-century gardeners is quite lengthy, offering many possible combinations for a kitchen garden. The herbs treated in these pages are documented and cross-referenced from books and manuscripts. This work focuses on eighteenth-century American sources and the British and other European sources that can be shown to have influenced them. Those herbal traditions are also traced back to earlier origins and forward into the nineteenth century.

From the pens and pages of early herbalists, gardeners, medical practitioners, cooks, travelers, and general householders you can piece together images of their gardens and the rationale behind their plant choices. Each plant in the eighteenth-century garden had a purpose—for culinary use, medicine, dyestuff, scent, beauty, superstition—and each plant, a story. Such stories are assembled here, fact and fancy that captured the interest of the historical gardener and still capture our attention today.

Books were costly then. The primary sources pivotal to our research were mostly found in prosperous households. Ideas from those venerable volumes were communicated to a broader audience through modest popular publications and shared between individual gardeners, cooks, and medical practitioners.

Early European herbals give insight into traditional usage. Plants known and used in European homelands were undoubtedly at the top of gardeners' wish lists. The volumes of Gerard, Parkinson, and Culpeper were influential on both sides of the Atlantic. In 1738 an unnamed South Carolinian copied herbal "virtues" from Gerard's 1633 *Herball or Generall Historie of Plantes* into a household memorandum book. Gerard had relied heavily on Dodoens' 1578 *Niewe Herball or Historie of Plants*, adding newly introduced American plants. The more affordable writings of Nicholas Culpeper were first published in 1633 and remained in print through two centuries, with few changes in later American editions.[1]

European herbals remained in high esteem in America through the eighteenth and nineteenth centuries, as did Old World garden calendars (particularly those of Miller, Bradley, and Mawe). New American garden calendars combined advice regarding Old World herbs with methods for New World gardens, adding newly discovered native plants. Gardening calendars and dictionaries presented an extensive catalog of desirable plants. Comparing Squibb's 1787 *Gardener's Calendar for South Carolina, Georgia and North-Carolina* against Mawe's 1773 English *Every Man His Own Gardener* reveals that

Squibb copied from Mawe, amending instructions for individual herbs to accommodate differences in growing conditions between England and the American South. Squibb candidly acknowledged following "the method and style of a justly admired European Gardener" but failed to name Mawe as that source.

Information from the early herbals continued to appear, even in popular medical sources. Early Americans sought advice from domestic medicine guides (see Buchan, Wesley, Tennent, Thomson, and Gunn) as well as volumes meant for the professional practitioner (as Quincy's *Complete English Dispensatory* and *The Edinburgh Dispensatory*). William Buchan's *Domestic Medicine* was published in 1769 with numerous subsequent editions in Britain and America. Popular on both sides of the Atlantic, *Domestic Medicine* continued to be published in America until 1913. Buchan's popularity in eighteenth-century American frontier communities is documented in the medical memorandum books of North Carolinians William Lenoir and Rachel Allen as well as John Thompson in Tennessee.

Volumes of household advice for the English "country housewife" reflect domestic traditions likely to have been transplanted to America in the early days of settlement. Cookery books shed light on Old World herbs in general use in Europe, those introduced into the American colonies, and New World plants of particular interest. Eliza Smith's *Compleat Housewife* was the first English cookery book published in America (Williamsburg, Virginia, 1742). Virginia household inventories and store ledgers indicate Hannah Glasse's *Art of Cookery* was the most popular English cookery book in America. Other imported volumes included those by Smith, Raffald, Bradley, Moxon, and even earlier works by May and Evelyn. New American cookbooks mingled many old culinary traditions with new and at times exotic receipts.[2]

Travelers' and settlers' journals yield fascinating details on culinary and medicinal herbs. Individual interests of early Americans are revealed in these personal memoranda. As we traced origins for particular bits of botanical history we attempted to make sense of connections between New World sources and Old World antecedents. Two examples of intricate connections are detailed in the preceding paragraphs: Remedies from English physician William Buchan's *Domestic Medicine* peppered American sources; receipts from Hannah Glasse and Eliza Smith showed up in numerous subsequent publications and personal receipt collections.[3]

Early authors copied, abridged, and compiled extensively from their predecessors—generally without acknowledging sources. In their day neither the ethics nor the legality of such derivative works was questioned or disparaged as plagiarism.

After-wort[4]

The new rage of the nineteenth century, the tomato, was unlikely to have been found in an eighteenth-century garden. Three centuries elapsed from Europeans' early-sixteenth-century discovery of the tomato in Central and South America until this fruit became widely enjoyed at English and North American tables. Seventeenth-century lore suggested the fruits "provoke a man to Lust," hence the name "love apple." Another common name, "wolf peach," derives from the Latin

fig. 58. Love Apple (Tomato)

Elizabeth Blackwell, *A Curious Herbal*, 1739

lycopersicom. In the early decades of the nineteenth century, receipts for tomatoes became ubiquitous in American cookery books and personal household memoranda.[5]

Tomato, Love Apple, Wolf Peach, *Solanum lycopersicum* (syn. *Lycopersicom esculentum*)

While a few sophisticated American gardeners were growing the love apple by mid-eighteenth century, the "middling sort," in both England and America, were slower to devote space and energy to this plant of questionable safety and worth. Adventurous gardeners began growing this member of the nightshade family as an ornamental herb before bringing the fruit into the kitchen. In a 1774 North Carolina gardener's calendar we find tomatoes listed among springtime plants for the kitchen garden.[6]

> [In 1629] Wee only have [Love Apples] for curiosity in our Gardens, and for the amorous aspect or beauty of the fruit.... In the hot Countries ... they are much eaten of the people, to coole and quench the heate and thirst.... The [Love] Apples also boyled, or infused in oyle in the sunne, is thought to be good to cure the itch, assuredly it will allay the heate thereof.
>
> —Parkinson, *Paradisi*, 380

> The Italians and Spaniards eat these Apples, as we do Cucumbers, with Pepper, Oil, and Salt; and some eat them stewed in Sauces, &c. and in Soups they are now [1735] much used in England.... This Fruit gives an agreeable acid to the Soup, though there are some Persons who think them not wholesome.
>
> —P. Miller, *Gardeners Dictionary*

The budding interest in the tomato for soups first spread to affluent English colonists in America such as South Carolina lowcountry plantation mistress Harriott Pinckney Horry, who penned directions "To Keep Tomatoos [*sic*] for Winter use" into her personal cookery book dated 1770. Similar methods for a concentrated or dry tomato paste to store for wintertime soups were published in *The Carolina Housewife* seventy-five years later. Tomatoes were slower to gain culinary favor in western Europe. *The French Cook* offered one lone tomato receipt, "Love-Apple Sauce," in 1828. German gardeners and cooks were likewise slow to adopt this suspicious member of the nightshade family.[7]

> [1822] Tomatoes are used to flavour sauces and soups in Southern Europe. They impart a pleasant, sharpish flavour and a fine orange colour.... I have no idea why people are not attempting to cultivate this savoury fruit in Germany.
>
> —Von Rumohr, *Essence of Cookery*, 162–63

> Plant them near fences, along which they branch out, or stakes.
>
> —Gardiner and Hepburn, *American Gardener*, 60

> It climbs up very high, and would require bushy sticks. It bears a sort of apple about as big as a black walnut.... This fruit is used to thicken stews and soups.
>
> —Cobbett, *American Gardener*, 143

> Love Apples ... the fruit is used for soups and pickles.
>
> —Gardiner and Hepburn, *American Gardener*, 36

In contrast to German and French attitudes, Mary Randolph published an assortment of tomato receipts in her 1824 *Virginia Housewife*, including the new rage, tomato catsup, and related tomato marmalades, but also

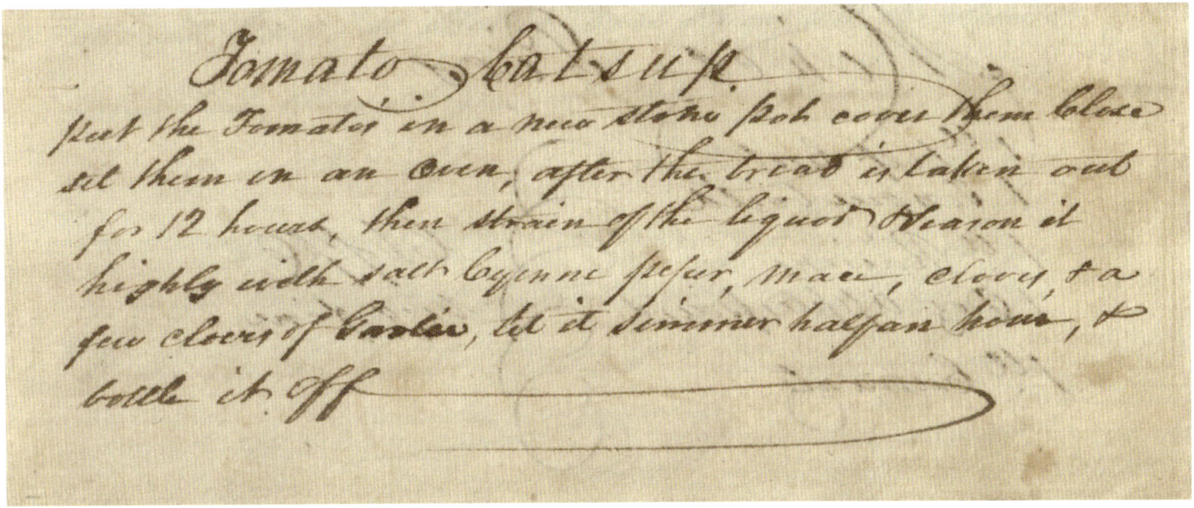

Tomato Catsup.
Put the Tomatos in a new stone pot cover them close
set them in an oven, after the bread is taken out
for 12 hours, then strain of the liquor & Season it
highly with salt Cayenne peper, mace, cloves, & a
few cloves of Garlic, let it simmer half an hour, &
bottle it off

FIG. 59. Tomato Catsup

Attributed to Anne and Jean Cameron, Recipe manuscript, North Carolina, c. 1834

tomatoes stewed, scalloped, and combined with okra. Indeed, in the early decades of the 1800s most American cookbooks included methods for stewing, baking, broiling, and frying tomatoes; for use in omelets or dressed raw as a salad; and in soups and sauces.[8]

Catsup, or ketchup, originally referred to a seasoning mixture based on the liquid of pickled shellfish, mushrooms, or walnuts with a combination of spices. Finally, the turn of the nineteenth century found all sorts of cooks penning directions for the popular new tomato catsup into personal commonplace books, as the piedmont North Carolina merchant family receipt below.

Tomato Catsup
Put the Tomatos in a new stone pot cover them close set them in an oven, after the bread is taken out for 12 hours, then strain of [*sic*] the liquor & Season it highly with Cayenne peper, mace, cloves, & a few cloves of Garlic, let it simmer half an hour, & bottle it off.

—Cameron, "A Friend to the Memory, Recipes"

Not the End

Where do we place the period to mark the end of this study? Our research has focused on identifying those plants most useful, most commonly mentioned, and most likely to have been grown in eighteenth-century southern and middle American gardens. But where do we draw the line? New garden plants continued to be introduced throughout this period. Tastes evolved. Medical opinions altered. Appreciation grew for the benefits to be gained from wild species. There seems no end to the list of native and naturalized plants of benefit to early Americans, especially to backcountry settlers.

This work has also focused on the research itself—methods, uncovering sources, and the sheer delight in discovery. What treasures may be awaiting *your* discovery?

It is now your turn to envision a personal garden and, if possible, to plant that garden. Perhaps you have been inspired to new culinary adventures. Perhaps you have cultivated an interest in native plants. Perhaps you will choose to delve more deeply into historical gardens and plant lore. Whatever your interests, we extend to you the challenge of the *Curious Garden*—beginning with the rich sources identified here and falling into the fascinating rabbit hole of research. We hope to have shared with you the seeds of Curiosity.

FIG. 60. An Herb Woman Gathering Simples

Thomas Green, *The Universal Herbal*, 1823 (detail)

ACKNOWLEDGMENTS

Two curious little girls exploring local woods and fields, separated by forty miles and near half that many years, were too busy to imagine future adult careers seriously studying those same woods and fields. The forest treasures we collected inspired creativity and fanned curiosity about the long-ago peoples who carved a life from that wilderness, changing the native landscape forever. Childhood games matured, as did the girls, along the way attracting like-minded friends and valued colleagues. Bound by a spirit of inquiry—scientific, environmental, historical, and gastronomic—those fellow woods wanderers and gardeners inspired and contributed to this book.

The greatest adventures in this realm accumulated during our decades as program specialists at the Schiele Museum of Natural History in Gastonia, North Carolina. Experimenters and collectors of the minutiae of life on the early southern frontier—historical interpreters, gardeners, cooks, craftspersons, and tellers of tales—have contributed more to research and field testing than they may ever realize. Those Living History Folk were not just casual museum volunteers, but were intimately involved in developing the program with an eye toward historical detail. We hope this volume will stir fond memories in those intrepid individuals who participated during four decades of Eighteenth-Century Backcountry Lifeways Studies events and workshops at the museum. To our Backcountry family: Well done. Huzzah!

So many museum colleagues encouraged and contributed to the projects that coalesced into this book. We feel so very fortunate to have belonged to an incredible staff who shared a bubbling stewpot of exciting ideas. More than a few fellow Schiele folks were adventurous enough to share our actual stewpots over an open fire, flavored with curious herbs from field, forest, and garden. Thank you, dear friends and partners, for support along the way. We would love to share a persimmon pudding or a sorrel posset with you all again.

A heartfelt thank-you to our initial peer readers, who represent a wide range of interests and specialized skills. Jim Daniel, Carolyn Dilda, Kim Niedert, and Claire Pittman reviewed our manuscript from widely and wildly varied points of view. We are so very grateful for their expertise and the hours they spent on their thoughtful analyses.

Anonymous peer reviewers (one American, another British) from the University of Georgia Press further shaped this project through a more focused perspective and constructive suggestions. Each asked insightful questions leading us on interesting side trips back into period sources. We thank you.

We are also deeply appreciative of the expertise of Dr. Larry Mellichamp, Professor of Botany Emeritus,

University of North Carolina at Charlotte, in confirming today's taxonomic nomenclature for the historic plants whenever possible. His precise attention to detail, good-natured patience with our questions, and generous gifts of time were essential to finalizing this work. Any errors that may remain are certainly ours.

Mary Glenn Krause, research assistant extraordinaire, when not tracking down botanical illustrations for this book, has picked enormous quantities of violet flowers and bravely tested centuries-old recipes.

The Museum of Early Southern Decorative Arts (MESDA) and Old Salem Museums and Gardens generously shared time and expertise, helping us identify several fine illustrations from their extensive collections. Contributions to this project by MESDA and Old Salem certainly do not begin there but ensue from decades of cooperation and sharing through workshops, conferences, publications, site visits, and informal staff collaboration. Likewise, the members of the Historic Cooking Guild of the Catawba Valley are always ready to share experiments with obscure ingredients and puzzling receipts. What delicious partnerships!

Unsung heroes, the Gaston County Public Library reference staff, have contributed so many small but hugely important services to this project—from years of procuring obscure books on loan to assisting with high-resolution scans.

A huge thank-you to our acquisitions editor, Patrick Allen, a steady source of encouragement and support. Patrick guided us through the hedgerows and brush piles of technology and image selection with an experienced hand and a touch of humor. A debt of gratitude to Jon Davies for clear, no-nonsense guidance going forward through the challenges of editing, design, and production. We hit the jackpot with Mindy Conner, our savvy copyeditor, whose mutual interests allowed her to understand and respect our goals and methods, far beyond spelling, grammar, and style. The University of Georgia Press staff members have been unwaveringly courteous and efficient, cultivating this garden from seed to harvest. Along the way, each one worked hand in hand with us as people, not just a product.

We began these acknowledgments with two little girls exploring the woods and fields; now we "girls at heart" recognize our greatest gratitude is owed to our families. From our grandparents and parents down to the present generation we each have been encouraged to explore the wild and grow a connection to the land, especially the wondrous world of plants. Much love and heartfelt thanks go to our cheerleader Rika Simmons; our hardy band of foragers and tasters, Susan, Ken, Mary Glenn, and Talley Krause, Meg Moss, Nathan and Adam Sears; and most especially to Fred Moss, who encouraged us to persevere!

GLOSSARY

Ague (n.) A fever; malaria; intermittent fevers.

Anthelmintic (n.) A deworming compound.

Bell (v.) To develop into seed heads (as hops).

Blanch (v.) To make a plant pale or white by covering to exclude light (as cardoon or celery); to remove skin from a nut or vegetable after briefly submerging it in boiling water (as almonds or peach kernels).

Blow (v.) To bloom.

Carmenative (n. or adj.) Relieving flatulence.

Cordial (adj.) Invigorating, stimulating the heart. (n.) Liqueur; tonic; sweet herb or fruit flavored drink.

Decoction (n.) Medicine or beverage prepared by boiling plant material in a liquid.

Drachm (n.) One-sixteenth of an ounce (weight); or simply a small quantity.

Empiric (n.) Medical practitioner without academic training; quack.

Extract (n.) Herbal essence concentrated in water or alcohol by decoction or infusion.

Flour, floure; flower (n.) Alternative spellings for "flower" (blossom) and for ground grains (flour or meal). Look to context for meaning.

Flux (n.) Diarrhea; excessive discharge of blood, excrement, or other fluid.

Fomentation (n.) A decoction meant for external application.

Galenic or humoral theory (n.) The idea that health can be attained through balancing the four humors (bodily fluids): blood, phlegm, choler or yellow bile, and melancholy or black bile. The four perceived qualities, *hotness*, *coldness*, *dryness*, and *moistness*, were considered when choosing an herbal food or medicine for restoring balance.

Gallipot (n.) Earthenware container, generally with glazed interior.

Hysteria (n.) Thought to be due to a disturbance of the uterus and its functions. . . . Former names for the disease were "vapours" and "hysteric(al) passion" (*OED*).

Infusion (n.) A beverage or remedy prepared by steeping or soaking plant material, usually in a hot liquid.

Melancholy (n.) Depression; sadness.

Middle colonies or mid-Atlantic (n.) Pennsylvania, Delaware, New Jersey, and New York.

Morphew (n.) Blemish on the skin; any of a variety of skin diseases.

Offal (n.) Entrails; discarded bits.

Physic, physick (n.) Medicine.

Physical (adj.) Medical.

Plantation (n.) Land where crops or garden plants are planted; may encompass a few square feet to hundreds of acres; a colony of emigrants (plants or people).

Receipt (n.) Recipe.

Rectified spirits (n.) Strong alcoholic liquid obtained by repeatedly distilling a fermented liquid. For example: "rectified spirits of wine" is otherwise known as brandy.

Rennet (n.) Lining from a calf's stomach containing the enzyme chymosin that curdles milk, the initial step in cheese making. Certain plants contain chemicals that similarly curdle milk, and serve as a vegetable rennet.

Signature; Doctrine of Signatures (n.) The appearance (as shape or color) of a plant believed to indicate its medical uses.

Simple (n.) A medicinal herb.

Southern colonies, or the South (n.) Georgia, South Carolina, North Carolina, Virginia, and Maryland.

Sweetmeat (n.) Confection; preserved, candied, or sugared fruits, nuts, or herbs.

Tincture (n.) Alcohol-based herbal extract.

Treacle (n.) Treacle in this context often refers to a cure-all medicine; otherwise "treacle" refers to a thick sweetener or syrup such as molasses.

Truck (n.) Market produce; garden stuff. (v.) To transport.

Wort (n.) A plant used medicinally or as food.

NOTES

Preface

1. Moss, *Backcountry Herbal.*

Introducing the Kitchen Garden

1. Fries, *Records of the Moravians*, 2: 575. In Baker's *Sons of a Trackless Forest* (333–34) we find gentlemen farmers on the Illinois frontier (James Alexander and George Morgan) seeking kitchen garden seeds from a Philadelphia source.

2. As quoted in Eleanour Sinclair Rohde, *A Garden of Herbs*, 280.

The Herbal

1. Tusser, *Five Hundred Points of Good Husbandry*, 124; unknown author, Herbal and household book MS (1738, South Carolina).

2. "Yesking"—belching, vomiting; "hicket"—hiccup.

3. Baker, *Sons of a Trackless Forest*. Ordered by James Alexander, 334.

4. P. Miller, *Gardeners Dictionary*, n.p.

5. Pre-1744 Virginia MS, 31, 32; Rutledge, *Carolina Housewife*, 97–98; M. Randolph, *Virginia Housewife*, 127.

6. Carter, *Frugal Housewife*, 59, 62, 96, 98.

7. Eringo, also called sea holly, *Eryngium maritimum*. The root was candied for a sweetmeat.

8. Von Rumohr, *Essence of Cookery*, 143.

9. J. Lawson, *New Voyage to Carolina*, 83; Kalm, *Travels*, 1: 89–90.

10. M. Randolph, *Virginia Housewife*, 121–22; Bryan, *Kentucky Housewife*, 203.

11. Thomson, *Family Physician*, 383.

12. "Taken from Dr. Salmons New London Dispensatory."

13. Squibb, *Gardener's Calendar*, 32; Bryan, *Kentucky Housewife*, 42–43, 84, 90.

14. Quincy, *English Dispensatory,* 86.

15. Unknown author, Herbal and household book (1738, South Carolina); Lenoir, Medical memorandum; pre-1744 Virginia MS, 41, 42; Culpeper, *English Physician*, 39.

16. Evelyn, *Aceteria*, 145.

17. The pre-1744 Virginia manuscript at MESDA is an interesting example.

18. Evelyn, *Acetaria* 158; Quincy, *English Dipsensatory*, 99.

19. J. Lawson, *New Voyage to Carolina*, 84; Pinckney, *Thomas and Elizabeth Lamboll*; Baker, *Sons of a Trackless Forest*, 334.

20. J. Randolph, *A Treatise on Gardening*, 24; Phillips, *History of Cultivated Vegetables*, 142; W. Lawson, *Country House-wives Garden*, 78.

21. Allen, Medical memorandum, n.d., Allen Family Papers.

22. Gerard, *Herball*, 1034; M'Mahon, *American Gardener's Calendar*, 321.

23. Fries, *Records of the Moravians*, 2: 574.

24. Fries, *Records of the Moravians*, 1: 164; Doddridge, *Notes on the Settlement*, 75.

25. J. Lawson, *New Voyage to Carolina*, 82; Fries, *Records of the Moravians;* Pinckney, *Thomas and Elizabeth Lamboll.*

26. Culpeper, *English Physician*, 35, 92.

27. Gerard, *Herball*, 807.

28. *Edinburgh Dispensatory*, 176; Gardiner and Hepburn, *American Gardener*, 35.

29. J. Lawson, *New Voyage to Carolina*, 83; J. Randolph, *Treatise on Gardening*, 23.

30. *Edinburgh Dispensatory*, 128.

31. J. Lawson, *New Voyage to Carolina*, 83; Fries, *Records of the Moravians*, 2: 573.

32. Gerard, *Herball*, 653.

33. M. Randolph, *Virginia Housewife*; Cameron, "A Friend to the Memory, Recipes."

34. Allen, Medical memorandum; Wesley, *Primitive Remedies*, 55, 93, 119; Lenoir, Medical memorandum, 1; Horry, Household book (1770, South Carolina).

35. W. Lawson, *Country House-wives Garden*, 80; Fries, *Records of the Moravians*, 2: 564.

36. Fries, *Records of the Moravians*, 2: 563; Lenoir, Medical memorandum; P. Miller, *Gardeners Kalendar*, 113; Tusser, *Five Hundred Points of Good Husbandry*, 136, 153.

37. P. Miller, *Gardeners Kalendar*, 114; Evelyn, *Aceteria*, 150.

38. Unknown author, Herbal and household book (1738, South Carolina).

39. Fries, *Records of the Moravians*, 2: 554; Tobler's 1759 *South Carolina Almanack*, South Carolina Historical Society, Charleston; Lenoir, Medical memorandum.

40. Gerard, *Herball*, 511; Thompson, Commonplace book (1803–11, Tennessee), 110. Several unrelated skin diseases were called St. Anthony's fire. The fiery rash of erysipelas, a streptococcal skin infection, was particularly feared in those days before antibiotics.

41. J. Randolph, *Treatise on Gardening*, 31; J. Lawson, *New Voyage to Carolina*, 83; Fries, *Records of the Moravians*, 2: 574.

42. J. Lawson, *New Voyage to Carolina*, 83; Phillips, *History of Cultivated Vegetables*, 1: 326; W. Lawson, *Country House-wives Garden*, 81; Carter, *Frugal Housewife*, 66; Rutledge, *Carolina Housewife*, 48–49.

43. The ancient Doctrine of Signatures held that a sign or signature indicated a plant's efficacy. For example: a heart-shaped leaf (as a woodsorrel leaflet) suggests use in heart complaints; stripping elder bark upward for tea to cause vomiting or downward when preparing a laxative; and the milky juice of some lettuces suggesting benefit to a nursing mother. See Quincy, *English Dispensatory*, 217.

44. Fries, *Records of the Moravians*, 2: 569; J. Lawson, *New Voyage to Carolina*, 83.

45. Loose paper with circa 1700 unidentified cookbook, in Harbury, *Virginia's Cooking Dynasty*, 280.

46. Glasse, *Art of Cookery*, 183–84.

47. Gerard, *Herball*, 751.

48. Von Rumohr, *Essence of Cookery*, 162; pre-1744 Virginia MS.

49. Fries, *Records of the Moravians*, 2: 574; J. Randolph, *Treatise on Gardening*, 34; Mawe, *Every Man His Own Gardener*, 479.

50. Rutledge, *Carolina Housewife*, 83.

51. Parkinson, *Paradisi*, 280–81.

52. M. Randolph, *Virginia Housewife*, 114.

53. Phillips, *History of Cultivated Vegetables*, 2: 3.

54. Phillips, *History of Cultivated Vegetables*, 2: 16; Evelyn, *Acetaria*, 156.

55. Phillips, *History of Cultivated Vegetables*, 2: 33.

56. *Edinburgh Dispensatory*, 168; Gerard, *Herball*, 171.

57. Tennent, *Every Man His Own Doctor*, 52, 53; Lenoir, Medical memorandum, 50.

58. P. Miller, *Gardeners Dictionary*; Squibb, *Gardener's Calendar*, 32.

59. Gunn, *Domestic Medicine*, 375.

60. Thomson, *Family Physician*, 412.

61. Tennent, *Every Man His Own Doctor*, 11, 12,14, 29, 50. Tennent described "vapours" or "hysterick fits" as "a great Heaviness, and Dejection of Spirit. . . . She has no Relish for any thing, but is continually out of Humour, she knows not why, and out of Order, she knows not where" (47–48); in

other words, anxiety or depression. Note the feminine pronoun, declaring these to be female infirmities.

62. Pinckney, *Letterbook*, 113; Squibb, *Gardener's Calendar*, 60.

63. B. Barton, *Cullen's Treatise*, 2: 149.

64. Bryan, *Kentucky Housewife*, 388.

65. W. Lawson, *Country House-wives Garden*, 82; Fries, *Records of the Moravians*, 2: 575; Von Rumohr, *Essence of Cookery*, 163. Hungary water, a popular cordial distilled from rosemary, was said to have been named for a queen of Hungary (*OED*).

66. Parkinson, *Paradisi*, 426.

67. P. Miller suggested that the name "herb of grace" comes from the practice of sprinkling holy water with rue. And in *Richard II*, Shakespeare introduces herb of grace as emblematic of sorrow, "in the remembrance of a weeping Queene"; Phillips, *History of Cultivated Vegetables*, 2: 142; Gunn, *Domestic Medicine*, 385.

68. Thomson, *Family Physician*, 417.

69. Parkinson, *Paradisi*, 167; Culpeper, *English Physician*, 295; Gerard, *Herball*, 152; Eliza Lucas Pinckney, Household book, 13.

70. Kalm, *Travels*, 1: 294; Bradley, *Survey of Ancient Husbandry and Gardening*, 129.

71. Thomson, *Family Physician*, 417; Phillips, *History of Cultivated Vegetables*, 2: 163.

72. Thomson, *Family Physician,* 417; Simmons, *American Cookery*, 43; Gunn, *Domestic Medicine*, 385.

73. W. Lawson, *Country House-wives Garden*, 83.

74. Gerard, *Herball*, 396; Bryan, *Kentucky Housewife*, 206.

75. Thomson, *Family Physician*, 425.

76. Culpeper, *English Physician*, 317; Gerard, *Herball*, 1107.

77. Gerard, *Herball*, 330; Cobbett, *American Gardener*, 143.

78. P. Miller, *Gardeners Dictionary*; Gunn, *Domestic Medicine*, 385; Cobbett, *American Gardener*, 144.

79. Gerard, *Herball*, 249.

80. Culpeper, *English Physician*, 333.

81. Dodoens (*Niewe Herball*, 522) suggested wearing this charm tied about the neck or arm.

82. Parkinson, *Paradisi*, 281.

83. Gerard, *Herball*, 852–53.

84. Worm fever was a disease caused by parasitic flatworms.

85. Byrd, *Histories*, 178.

86. Lawson, *New Voyage to Carolina*, 83; Phillips, *History of Cultivated Vegetables*, 1: 297.

87. P. Miller, *Gardeners Dictionary*, n.p.

88. Thompson, Commonplace book, Cameron Family Papers; Johnson, *Every Man His Own Doctor*, 18.

89. Evelyn, *Acetaria*, 155; Cobbett, *American Gardener*, 132–33; *Edinburgh Dispensatory*, 283.

90. Kalm, *Travels*, 1: 64; Fries, *Records of the Moravians*, 2: 565.

91. Doddridge, *Notes on the Settlement*, 117; Lenoir, Medical memorandum; Wesley, *Primitive Remedies*, 111.

92. Rutledge, *Carolina Housewife*, 221; B. Barton, *Collections*, 2: 27–28.

93. J. Lawson, *New Voyage to Carolina*, 83.

94. Cleland, *New and Easy Method of Cookery*, 210.

95. Culpeper, *English Physician*, 123.

96. J. Lawson, *New Voyage to Carolina*, 109.

97. Kalm, *Travels*, 1: 39, 337.

98. Gerard, *Herball*, 1525.

99. Letter from Chapel Hill to Charlotte, North Carolina, Springs Family Papers, Southern Historical Collection, University of North Carolina, Chapel Hill.

100. Dried sassafras leaves are otherwise known as filé. In soups and gumbos okra also serves for both a thickening agent and flavor. Kalm, *Travels*, 1: 42.

101. *Edinburgh Dispensatory*, 292.

102. M'Mahon, *American Gardener's Calendar*, 269.

103. Fries, *Records of the Moravians*, 2: 564; Kalm, *Travels*, 1: 48; Michaux, "Travels 1801–1802," 241.

104. Fries, *Records of the Moravians*, 1: 99; Michaux, "Travels 1801–1802," 269.

105. J. Lawson, *New Voyage to Carolina*, 115; Michaux, "Travels 1801–1802," 298–99, 269.

106. See "jumbals" and "Shrewsbury cakes" in Rutledge, *Carolina Housewife*, 202; and "little cakes," unknown author, Herbal and household book, 1738, 1778–85, South Carolina.

107. Bradley, *Country Housewife*, 2: 70–71.

108. J. Lawson, *New Voyage to Carolina*, 105.

109. Kalm, *Travels*, 1: 161.

110. Kalm, *Travels*, 1: 165; J. Lawson, *New Voyage to Carolina*, 105.

111. J. Lawson, *New Voyage to Carolina*, 107.

112. Fries, *Records of the Moravians*, 2: 576.

A Salmagundi of Methods

1. Dods, *Cook and Housewife's Manual*, 225, 226.

2. Von Rumohr, *Essence of Cookery*, 97.

3. Fries, *Records of the Moravians*, 2: 573.

4. Approximately one-half to three quarters pound. The size of a penny loaf varied with the price of flour.

5. Elise Pinckney and Eliza Lucas Pinckney, "Letters of Eliza Lucas Pinckney," 144; peach water used as flavoring in Rutledge's *Carolina Housewife* 138, 142, 190, 194, 196, 203–4, etc.

6. Dods, *Cook and Housewife's Manual*, 456.

7. Then as now, saffron was extremely costly; sugar was much more dear than now.

8. Dods, *Cook and Housewife's Manual*, 256; Elizabeth Isabella Purviance, Household book.

9. Buchan, *Domestic Medicine*, 704.

10. May, *Accomplisht Cook*, 139.

11. Dods, *Cook and Housewife's Manual*, 440.

12. Pre-1744 Virginia cookbook.

13. Bryan, *Kentucky Housewife*, 436.

Miscellaneous Gardening Theories and Practices

1. Fries, *Records of the Moravians*, 4: 1505.

2. Brooks, Groover, and Smith, *Living on the Edge*, 181.

Epilogue

1. Unknown author, Herbal and household book, 1738 (1778–85, South Carolina lowcountry).

2. Karen Hess, ed., "Historical Notes and Commentories," in M. Randolph, *Virginia Housewife*, xiv; note from Carson, *Colonial Virginia Cookery*.

3. Those two cookery classics had drawn heavily from a less successful mid-eighteenth-century volume, *The Whole Duty of a Woman; or an Infallible Guide to the Fair Sex* (London, 1737).

4. "After-wort" is our own play on words.

5. Mateo Aleman, *The Rogue* (London, 1622) (as per *OED*).

6. "A Gardener's Kalendar for North Carolina," in *Wells Register* (Charleston: Robert Wells, 1774).

7. Horry, *Colonial Plantation Cookbook*, 89; Ude, *French Cook*, 40.

8. Bryan, *Kentucky Housewife*, 216–18; Rutledge, *Carolina Housewife*, 102–6.

BIBLIOGRAPHY

Primary Sources

Allen, Rachel Stout. Medical memorandum (c. late 1700s). Piedmont North Carolina (Guilford Co.). Manuscript in Allen Family Papers, North Carolina State Archives, Raleigh.

Anburey, Thomas. *Travels through the Interior Parts of America*. London, 1789. Reprint. New York: New York Times and Arno Press, 1960.

Barton, Benjamin Smith. *Collections for an Essay towards a Materia Medica of the United States*. Philadelphia, 1810.

———. *Professor Cullen's Treatise of the Materia Medica*. Philadelphia, 1812.

Barton, William P. C. *Vegetable Materia Medica of the United States*. Philadelphia, 1825.

Bartram, William. *Travels* (1773–78). Philadelphia, 1791. Reprint. New York: Dover, 1955.

Besler, Basilius. *Besler's Book of Flowers & Plants: 73 Full-Color Plates from Hortus Eystettensis*. 1613. Mineola, N.Y.: Dover, 2007.

Beverly, Robert. *The History of Virginia, in Four Parts*. London, 1722; Virginia, 1855.

Blackwell, Elizabeth. *A Curious Herbal, containing five hundred Cuts, of the most useful Plants, which are now used in the practice of Physick.* Vol. 1. London, 1782. Science, Industry and Business Library: General Collection, New York Public Library.

Blount, Jacob. Pocket memorandum (1764–67). Eastern North Carolina. Manuscript in North Carolina State Archives, Raleigh.

Bradley, Richard. *The Country Housewife and Lady's Director*. 1736. London: Prospect Books, 1980.

———. *The Gentleman and Gardener's Kalendar*. London, 1720. Gale ECCO Print Edition, ESTCID: T040724. Reproduction from British Library.

———. *Survey of Ancient Husbandry and Gardening*. 1725. Special Collections, USDA National Agricultural Library.

Bryan, Lettice. *The Kentucky Housewife*. 1839. Columbia: University of South Carolina Press, 1991.

Buchan, William. *Domestic Medicine; or, a Treatise on the Prevention and Cure of Diseases by Regimen and Simple Medicines*. 3rd ed. London, 1774. [This edition is the one referenced in the text unless otherwise noted.]

———. *Every Man His Own Doctor; or, a Treatise on the Prevention and Cure of Diseases, by Regimen and Simple Medicines. To Which Is Added a Treatise on the Materia Medica*. New Haven: Nathan Whiting, 1816. [Very similar to 1794 *Domestic Medicine*, with additions including Materia Medica extracted from Thacher and Barton.]

Byrd, William. *Histories of the Dividing Line betwixt Virginia and North Carolina*. 1728. Gloucester, Mass.: Peter Smith, 1984.

Cameron. "A Friend to the Memory, Recipes, Mrs. R. Cameron, Fairntosh, Orange Co. No. Carolina. October the 15th 1816." Recipes attributed to Annie Cameron, circa 1834. Miscellaneous cures. Manuscript in Cameron Family Papers, Southern Historical Collection, University of North Carolina, Chapel Hill.

Carson, Jane. *Colonial Virginia Cookery*. Williamsburg, Va.: Colonial Williamsburg, 1968.

Carter, Susannah. *The Frugal Housewife*. London and Boston, 1772. Reprint. Garden City, N.Y.: Doubleday, 1976.

Castiglioni, Luigi. *Viaggio: Travels in the United States of North America 1785–1787 with Observations on the Most Useful Plants in the United States*. Trans. from the Italian by Antonio Pace. Syracuse, N.Y.: Syracuse University Press, 1983.

Child, Mrs. [Lydia Maria]. *The American Frugal Housewife*. 1832. Boston: Carter, Hendee, and Co., 1980.

Cleland, Elizabeth. *A New and Easy Method of Cookery*. Edinburgh: C. Wright and Co., 1759. Gale ECCO Print Editions, 2010.

Cobbett, William. *The American Gardener*. 1828, 1836. Reprint. New York: Random House, 2003.

Collet, John Abraham. "A Map-Maker's View of Anson County in 1769." Ed. H. Roy Merrens and Herbert R. Paschal. *North Carolina Historical Review* 59, no. 3 (1982): 271–78.

Conner, Juliana Margaret. Diary. 1827. Typescript in Conner Papers, Southern Historical Collection, University of North Carolina, Chapel Hill.

Culpeper, Nicholas. *Complete Herbal and English Physician* . . . [and] *The British Florist*. 1826.

———. *The English Physician*. Enlarged. London, 1770.

Doddridge, Joseph. *Notes on the Settlement and Indian Wars of the Western Parts of Virginia and Pennsylvania from 1763 to 1783 . . . 1824*. Pittsburgh: John S. Titenour and William T. Lindsey, 1912.

Dodoens, Rembert. *A Niewe Herball or Historie of Plants*. Trans. Henry Lyte. France, 1578.

Dods, Mistress Margaret. *The Cook and Housewife's Manual*. 4th ed. 1829. Reprint. London: Rosters, 1988.

Edinburgh New Dispensatory. 3rd ed. Edinburgh, 1791. Based on William Lewis' *New Dispensatory*.

Ellis, William. *The Country Housewife's Family Companion*. 1750. Reprint. Devon, U.K.: Prospect Books, 2000.

Evelyn, John. *Directions for the Gardiner and Other Horticultural Advice*. 1686. Bound with *Kalendarium Hortense; or the Gard'ners Almanac*, 1664; and *Acetaria: A Discourse of Sallets*, 1699. Oxford: Oxford University Press, 2009.

Folger, Alfred M. *The Family Physician*. Stokes County, N.C. Spartanburg, S.C., 1845.

Fries, Adelaide L., ed. *Records of the Moravians in North Carolina*. Vols. 1–6. Raleigh, N.C.: Department of Archives and History, 1968.

Gale's North Carolina Almanac for 1830. North Carolina State Archives, Raleigh, PC 89.11.

Gardiner, John, and David Hepburn. *The American Gardener; to Which Is Added a Treatise on Gardening*. Washington City, 1826. Google Books VQD Harvard Book Store.

Gerard, John. *The Herball or Generall Historie of Plantes*. 1633, 1636. Reprint. New York: Dover, 1975.

Glasse, Hannah. *The Art of Cookery Made Plain and Easy*. 1747. Reprint. London: Prospect Books, 1983.

———. *The Art of Cookery Made Plain and Easy*. American edition, 1804. Reprint. Bedford, Mass.: Applewood Books, 1997.

Goelet [Elizabeth?]. Household book. 1756. North Carolina. Manuscript in Goelet-Buncombe Papers, Southern Historical Collection, University of North Carolina, Chapel Hill.

Green, Thomas. *The Universal Herbalist; or Botanical, Medical, and Agricultural Dictionary*. Vols. 1 and 2. London, 1823.

Gunn, John C. *Domestic Medicine*. 1830. Knoxville: University of Tennessee Press, 1986.

Horry, Harriott Pinckney. *A Colonial Plantation Cookbook: The Receipt Book of Harriott Pinckney Horry, 1770*. Ed. Richard Hooker. Columbia: University of South Carolina Press, 1984.

———. Household book. Begun 1770. South Carolina lowcountry. Manuscript in Pinckney Papers, South Carolina Historical Society, Charleston.

Howard, Horton. *Howard's Botanical Medicine*. Columbus, Ohio: for the author, 1836.

Instructions of All Kinds of Cookery and Pastry for Young Women. By a Friend of Cookery. Geppingen, 1787. Trans. from the German in *The Raised Hearth*, ed. Joanna Roberts. Winston-Salem, N.C.: Old Salem Museums and Gardens, 2017.

Johnson, Thomas. *Every Man His Own Doctor*. Salisbury, N.C., 1798.

Kalm, Peter. *Travels in North America* [Swedish]. The English Version of 1770. Vols. 1 and 2. New York: Dover, 1966.

La Varenne, François Pierre. *The French Cook*. "Englished by I.D.G. 1653." Lewes, East Sussex, U.K.: Southover Press, 2001.

Lawson, John. *A New Voyage to Carolina*. London, 1709. Reprint. Chapel Hill: University of North Carolina Press, 1967.

Lawson, William. *The Country House-wives Garden*. Bound with *A New Orchard and Garden*. London, 1660.

Lenoir, William. Medical memorandum. 1801–39. Western North Carolina. Manuscript in Lenoir Family Papers, vol. 68, Southern Historical Collection, University of North Carolina, Chapel Hill.

Loffler, Friederike Luise. *The Economical Handbook for Housewives*. Stuttgard, 1795. Trans. from the German in *The Raised Hearth*, ed. Joanna Roberts. Winston-Salem, N.C.: Old Salem Museums and Gardens, 2017.

Martha Washington's Booke of Cookery. By an unknown hand. Circa 1600s. Ed. Karen Hess. New York: Columbia University Press, 1981.

Mawe, Thomas. *Every Man His Own Gardener*. London, 1773.

May, Robert. *The Accomplisht Cook; or, The Art & Mystery of Cookery*. London, 1685. Reprint. Lexington, Ky.: Filiquarian Publishing, 2010.

Merian, Maria Sibylla. *Erucarum Ortus Alimentum et Paradoxa Metamorphosis*. Amsterdam, 1718. *Merian's Antique Botanical Prints*. New York: Dover, 2005.

Michaux, François André. "Travels 1801–1802." Trans. from the French and ed. Reuben Gold Thwaites. In *Early Western Travels 1748–1846*. Vol. 3, 1906. New York: AMS Press, 1966.

Miller, Joseph. *Botanicum Officinale*. London, 1722.

Miller, Lewis. *Lewis Miller: Sketches and Chronicles*. Nineteenth century. York, Pa.: Historical Society of York County, 1966.

Miller, Philip. *The Gardeners Dictionary*. London, 1735, 1759.

———. *The Gardeners Kalendar*. London, 1759.

M'Mahon, Bernard. *The American Gardener's Calendar*. 1806.

Moreau de St. Méry's American Journey (1793–1719). Trans. from the French and ed. Kenneth Roberts and Anna M. Roberts. Garden City, N.Y.: Doubleday, 1947.

Moxon, Elisabeth. *English Housewifry*. Leeds: James Lister, 1742.

Parkinson, John. *Paradisi in Sole: Paradisus Terrestris*. 1629. Reprint. New York: Dover, 1976.

Passe, Crispin van de. *A Garden of Flowers: All 104 Engravings from The Hortus Floridus of 1614*. Mineola, N.Y.: Dover, 2002.

Phillips, Henry. *History of Cultivated Vegetables*. Vols. 1 and 2. London: Henry Colburn and Co., 1822.

Pinckney, Elise, and Eliza Lucas Pinckney. "Letters of Eliza Lucas Pinckney, 1768–1782." *South Carolina Historical Magazine* 76, no. 3 (1975).

Pinckney, Eliza Lucas. Household book. "Recipe Book No. 2." Begun 1756, South Carolina lowcountry. On deposit by the Colonial Dames of America in the State of South Carolina in the South Carolina Historical Society, Charleston.

———. *The Letterbook of Eliza Lucas Pinckney 1739–1762*. Ed. Elise Pinckney. Chapel Hill: University of North Carolina Press, 1972.

Pre-1744 Virginia cookbook. Manuscript and typescript. Old Salem Museums and Gardens, Winston-Salem, N.C.

Purviance, Elizabeth Isabella. Household book. Circa 1799. Baltimore, Md. Manuscript in Purviance-Courtaney Papers, Special Collections, Duke University, Durham, N.C.

Pyne, William Henry. *Picturesque Views of Rural Occupations in Early Nineteenth-Century England. England: 1802–1807*. Reprint. New York: Dover, 1977.

Quincy, John. *A Complete English Dispensatory*. London, 1736.

Raffald, Elizabeth. *The Experienced English Housekeeper*. 1769. Reprint. Lewes, East Sussex, U.K.: Southover Press, 1997.

Randolph, Jane (Bolling). "Jane Randolph her Cookery Book." 1743. In *Colonial Virginia's Cooking Dynasty*, ed. Katharine E. Harbury, pp. 285–421. Virginia Historical Society, Richmond. Columbia: University of South Carolina Press, 2004.

Randolph, John. See *A Treatise on Gardening* "By a native of this State."

Randolph, Mary. *The Virginia Housewife*. Washington, 1824. Reprint. Columbia: University of South Carolina Press, 1984.

Rundell, Maria Eliza ("By a Lady"). *A New System of Domestic Cookery; formed upon Principles of Economy; and Adapted to the Use of Private Families*. Boston, 1807; and London, 1810.

Rutledge, Sarah. *The Carolina Housewife*. 1847. Reprint. Columbia: University of South Carolina Press, 1980.

Sayer, Robert. *The Florist*. London, 1760.

Simmons, Amelia. *American Cookery*. 1796. Reprint. Green Farms, Conn.: Silverleaf Press, 1984.

Smith, E. *The Compleat Housewife; or, Accomplish'd Gentlewoman's Companion*. London, 1742.

Smyth, John Ferdinand Dalziel. *A Tour in the United States of America*. Vol. 1. London, 1784. Reprint. New York: New York Times and Arno Press, 1968.

The South Carolina and Georgia Almanack for the Year of Our Lord 1758. Charleston: John Tobler, 1758.

Squibb, Robert. *The Gardener's Calendar for South Carolina, Georgia and North-Carolina*. Charleston, 1787. Reprint. Athens: University of Georgia Press, 1980.

Tennent, John. *Every Man His Own Doctor; or The Poor Planter's Physician*. Williamsburg, 1736. Reprint. Williamsburg, Va.: Printing Office of Colonial Williamsburg, 1984.

Thompson, John. Commonplace book. 1803–11. Central Tennessee (Davidson County). MS. Special Collections, University of Tennessee, Knoxville.

Thomson, Alexander. *The Family Physician; or Domestic Medical Friend*. 1st American ed. New York, 1802.

A Treatise on Gardening. "By a native of this State." Attributed to John Randolph, Williamsburg, Va. Richmond, 1793. Reprint. Williamsburg, Va.: Printing Office of the Colonial Williamsburg Foundation. Also bound with Gardiner and Hepburn, *The American Gardener*, 1826.

Tusser, Thomas. *Five Hundred Points of Good Husbandry, together with A Book of Huswifery*. 1573, 1580, 1585. London, 1812.

Ude, Louis Eustache. *The French Cook*. 1828. Reprint. New York: Arco, 1978.

Unidentified cookbook. Circa 1700. Manuscript. Reprinted in *Colonial Virginia's Cooking Dynasty*, ed. Katharine E. Harbury, pp. 145–283. Virginia Historical Society, Richmond. Columbia: University of South Carolina Press, 2004.

Unknown author. Herbal and household book. 1738, 1778–85. South Carolina lowcountry. Manuscript in Preston-Davie Papers, Southern Historical Collection, University of North Carolina, Chapel Hill.

Valuable Secrets in Arts, Trades, &c. Selected from the Best Authors and Adapted to the Situation of the United States.

New York, 1816. Earlier editions included Dublin, 1778; and America, 1795.

Von Rumohr, Karl Friedrick. *The Essence of Cookery*. 1822. Trans. from the German by Barbara Yoemans. London: Prospect Books, 1993.

Wells Register. Charleston: Robert Wells, 1774.

Wesley, John. *Primitive Remedies*. [Originally *Primitive Physick*, 1776.] Santa Barbara: Woodbridge Press, 1975.

Winkler Family recipe. MS. Archives of the Moravian Church in America, Southern Province, Winston-Salem, N.C. Trans. from the German. Old Salem Inc., Winston-Salem, N.C.

Wolley, Hannah. *The Queen-like Closet*. 1672. Reprint. Charleston, S.C.: Bibliobazaar, 2006.

Among secondary sources you may find the following helpful.

Baker, Mark A. *Sons of a Trackless Forest: The Cumberland Long Hunters of the Eighteenth Century*. Franklin, Tenn.: Baker's Trace Publishing, 1997.

Brooks, Richard D., Mark D. Groover, and Samuel C. Smith. *Living on the Edge: The Archaeology of Cattle Raisers in the South Carolina Backcountry*. Savannah River Archaeological Research Papers 10. Occasional Papers of the Savannah River Archaeological Research Program, South Carolina Institute of Archaeology and Anthropology, University of South Carolina, 2000.

Burford, Tom. *Apples of North America*. Portland, Oreg.: Timber Press, 2013.

Bynum, Flora Ann L. *Cultivated Plants of the Wachovia Tract in North Carolina*. Pamphlet. Winston Salem: Old Salem, Ann L. Bynum, 1979.

———. *Old Salem Garden Guide*. Winston-Salem: Old Salem Inc., 1979.

Calhoun, Creighton Lee Jr. *Old Southern Apples*. Blacksburg, Va.: McDonald and Woodward, 1995.

From Seed to Flower: Philadelphia 1618–1876. Philadelphia: Pennsylvania Horticultural Society, 1976.

Greene, Wesley. *Vegetable Gardening the Colonial Williamsburg Way*. Colonial Williamsburg Foundation. New York: Rodale, 2012.

Hatch, Peter J. *"A Rich Spot of Earth": Thomas Jefferson's Revolutionary Garden at Monticello*. Thomas Jefferson Foundation. New Haven: Yale University Press, 2012.

Moss, Kay [K]. *A Backcountry Herbal*. Gastonia, N.C.: Schiele Museum, 1993.

Moss, Kay K. *Seeking the Historical Cook*. Columbia: University of South Carolina Press, 2013.

———. *Southern Folkmedicine, 1750–1820*. Columbia: University of South Carolina Press, 1999.

Moss, Kay, and Kathryn Hoffman. *The Backcountry Housewife*. Vol. 1: *A Study of Eighteenth Century Foods*. Gastonia, N.C.: Schiele Museum, 1985.

Moss, Kay K., with Ann Tippitt, Suzanne Simmons, Steve Watts, and Alan May. *Journey to the Piedmont Past*. Gastonia, N.C.: Schiele Museum, 1997.

Pinckney, Elise. *Thomas and Elizabeth Lamboll: Early Charleston Gardeners*. Charleston: Charleston Museum, 1969.

Rohde, Eleanour Sinclair. *A Garden of Herbs*. New York: Dover, 1969.

Stetson, Sarah Pattee. "American Garden Books Transplanted and Native, before 1807." *William and Mary Quarterly* 3, no. 3 (1946): 343–69.

Viola, Herman J., and Carolyn Margolis, eds. *Seeds of Change*. Washington and London: Smithsonian Institution Press, 1991.

Weaver, William Woys. *Heirloom Vegetable Gardening*. New York: Henry Holt, 1997.

Welch, William C. *Antique Roses for the South*. Dallas: Taylor Publishing Co., 1990.

ILLUSTRATION CREDITS

FIG. 1. Courtesy of Thomas Fisher Rare Book Library, University of Toronto, Canada.

FIG. 2. Collection of the Moravian Archives, Bethlehem, Pa.

FIG. 3. Collection of the Museum of Early Southern Decorative Arts, Old Salem Museums and Gardens; gift of Dale C. Couch.

FIG. 4. Collection of the Wachovia Historical Society; photograph courtesy of Old Salem Museums and Gardens.

FIG. 5. Collection of the Museum of Early Southern Decorative Arts, Old Salem Museums and Gardens.

FIG. 6. Courtesy of Thomas Fisher Rare Book Library, University of Toronto, Canada.

FIG. 7. Special Collections, USDA National Agricultural Library.

FIG. 8. Special Collections, USDA National Agricultural Library.

FIG. 9. Courtesy of Peter H. Raven Library, Missouri Botanical Garden.

FIG. 10. Courtesy of Science, Industry and Business Library: General Collection, New York Public Library Digital Collections.

FIG. 11. Image from Gerard, *The Herball or Generall Historie of Plants*, Dover Electronic Clip Art Series.

FIG. 12. Image from Gerard, *The Herball or Generall Historie of Plants*, Dover Electronic Clip Art Series.

FIG. 13. Courtesy of Getty Research Institute.

FIG. 14. Image from Gerard, *The Herball or Generall Historie of Plants*, Dover Electronic Clip Art Series.

FIG. 15. Image from Gerard, *The Herball or Generall Historie of Plants*, Dover Electronic Clip Art Series.

FIG. 16. Courtesy of Thomas Fisher Rare Book Library, University of Toronto, Canada.

FIG. 17. Image from the collection of the York County History Center, York, Pa., Lewis Miller, *Sketches and Chronicles*, Historical Society of York, Pennsylvania, 1966, page 73.

FIG. 18. Courtesy of Wellcome Library.

FIG. 19. Courtesy of University of California Digital Libraries.

FIG. 20. Image from the Biodiversity Heritage Library; contributed by Smithsonian Libraries.

FIG. 21. Image from W. H. Pyne, *Picturesque Views of Rural Occupations*, 1977, Dover Pictorial Archive Series.

FIG. 22. Image from Gerard, *The Herball or Generall Historie of Plants*, Dover Electronic Clip Art Series.

FIG. 23. Image from Gerard, *The Herball or Generall Historie of Plants*, Dover Electronic Clip Art Series.

FIG. 24. Courtesy of Science, Industry and Business Library: General Collection, New York Public Library Digital Collections.

FIG. 25. Special Collections, USDA National Agricultural Library.

FIG. 26. Courtesy of Peter H. Raven Library, Missouri Botanical Garden.

FIG. 27. Courtesy of Wellcome Library.

FIG. 28. Courtesy of the Medical Heritage Library; contributed by the U.S. National Library of Medicine.

FIG. 29. Courtesy of Peter H. Raven Library, Missouri Botanical Garden.

FIG. 30. Courtesy of Southern Historical Collection, Louis Round Wilson Special Collections Library, University of North Carolina at Chapel Hill.

FIG. 31. Kay Moss collection.

FIG. 32. Image from the California Digital Library; contributed by University of California Libraries.

FIG. 33. Courtesy of Science, Industry and Business Library: General Collection, New York Public Library Digital Collections.

FIG. 34. Courtesy of U.S. National Library of Medicine.

FIG. 35. Courtesy of Science, Industry and Business Library: General Collection, New York Public Library Digital Collections.

FIG. 36. Courtesy of Thomas Fisher Rare Book Library, University of Toronto, Canada.

FIG. 37. Image from Gerard, *The Herball or Generall Historie of Plants*, Dover Electronic Clip Art Series.

FIG. 38. Image from Gerard, *The Herball or Generall Historie of Plants*, Dover Electronic Clip Art Series.

FIG. 39. Courtesy of Collection of the Museum of Early Southern Decorative Arts, Old Salem Museums and Gardens.

FIG. 40. Courtesy of Science, Industry and Business Library: General Collection, New York Public Library Digital Collections.

FIG. 41. Image from Gerard, *The Herball or Generall Historie of Plants*, Dover Electronic Clip Art Series.

FIG. 42. Courtesy of Science, Industry and Business Library: General Collection, New York Public Library Digital Collections.

FIG. 43. Courtesy of Wellcome Library.

FIG. 44. Courtesy of Science, Industry and Business Library: General Collection, New York Public Library Digital Collections.

FIG. 45. Courtesy of Thomas Fisher Rare Book Library, University of Toronto, Canada.

FIG. 46. Courtesy of Science, Industry and Business Library: General Collection, New York Public Library Digital Collections.

FIG. 47. Image from Biodiversity Heritage Library; contributed by Smithsonian Libraries.

FIG. 48. Image from Biodiversity Heritage Library; contributed by Smithsonian Libraries.

FIG. 49. Courtesy of Irvin Department of Rare Books and Special Collections, Pope-Brown Watercolors, University of South Carolina Libraries, Columbia.

FIG. 50. Courtesy of Science, Industry and Business Library: General Collection, New York Public Library Digital Collections.

FIG. 51. From *The Backcountry Housewife* (1985), Teresa Armour, artist; courtesy of the Schiele Museum of Natural History and Planetarium Inc., Gastonia, N.C.

FIG. 52. Saucepan from the collection of Culture and Heritage Museums of York County, S.C. Drawing from *The Backcountry Housewife* (1985), Teresa Armour, artist; courtesy of the Schiele Museum of Natural History and Planetarium Inc., Gastonia, N.C.

FIG. 53. From *The Backcountry Housewife* (1985), Teresa Armour, artist; courtesy of the Schiele Museum of Natural History and Planetarium Inc., Gastonia, N.C.

FIG. 54. From *The Backcountry Housewife* (1985), Teresa Armour, artist; courtesy of the Schiele Museum of Natural History and Planetarium Inc., Gastonia, N.C.

FIG. 55. From *The Backcountry Housewife* (1985), Teresa Armour, artist; courtesy of the Schiele Museum of Natural History and Planetarium Inc., Gastonia, N.C.

FIG. 56. From *The Backcountry Housewife* (1985), Teresa Armour, artist; courtesy of the Schiele Museum of Natural History and Planetarium Inc., Gastonia, N.C.

FIG. 57. Courtesy of Syracuse University Press, Syracuse, N.Y.

FIG. 58. Courtesy of Science, Industry and Business Library: General Collection, New York Public Library Digital Collections.

FIG. 59. Courtesy of Louis Round Wilson Special Collections Library, Southern Historical Collection, University of North Carolina at Chapel Hill.

FIG. 60. Kay Moss collection.

INDEX OF SCIENTIFIC NAMES

INDEX

Bolded numbers indicate main species entry.